Soil and Water Engineering

Soil and Water Engineering

Editor

Dalmiro Martell

scitus
academics

Soil and Water Engineering
Edited by **Dalmiro Martell**

ISBN: 978-1-68117-706-9
Library of Congress Control Number: 2016958746

SCITUS Academics LLC,
616, Corporate Way, Suite 2, 4766,
Valley Cottage, NY 10989

www.scitusacademics.com

Notice

Printed in United States of America on Acid Free Paper

Preface

Soil can be the small skin that covers the ground; clean water is more valuable every day. As global populations grow and food demand increases in a community with a fixed water supply, efforts are underway to find the right way to create far more food with less water by improving irrigation methods as well as technologies, agricultural water management, crop types and water monitoring. The main challenges facing managers and water engineers today include securing water for people and food production: protecting vital ecosystems and addressing the uncertainty and variability of water in time and space. Modeling aspects in the field of soil and water engineering have added an innovative dimension to research innovation in all engineering sectors. They are often used to manage, develop and plan land and water resources, including analysis of soil and surface water quantity and quality parameters, food forecasting and control measures, optimal allocation and use of irrigation water.

Focusing on the soil/water interface and how to manage it for maximum productivity in the long-term sustainability and overall ecosystem health requirements, this book provides a balanced approach that protects both the design and management aspects of the development of water resources. It also addresses water control methods to prevent water damage and disasters.

This book will be an extensive guide for students, researchers and practitioners and water resources engineering professionals in general to learn about the latest technological advances and changes.

Table of Contents

1

CHAPTER

Potential Impacts of Food Production on Freshwater Availability Considering Water Sources

Shinjiro Yano [1,*], Naota Hanasaki [2], Norihiro Itsubo [3] and Taikan Oki [4]

[1] *Institute for Water Science, Suntory Global Innovation Center Limited, 8-1-1 Seikadai, Seika-cho, Soraku-gun, Kyoto 619-0284, Japan*

[2] *Center for Global Environmental Research, National Institute for Environmental Studies, 16-2 Onogawa, Tsukuba, Ibaraki 305-8506, Japan*

[3] *Faculty of Environmental Studies, Tokyo City University, 3-3-1 Ushikubo-nishi, Tsuzuki-ku, Yokohama, Kanagawa 224-8551, Japan*

[4] *Institute of Industrial Science, The University of Tokyo, 4-6-1 Meguro-ku, Komaba, Tokyo 153-8505, Japan*

ABSTRACT

We quantify the potential impacts of global food production on freshwater availability (water scarcity footprint; WSF) by applying the water unavailability factor (*fwua*) as a characterization factor and a global water resource model based on life cycle impact assessment (LCIA). Each water source, including rainfall, surface water, and groundwater, has a distinct *fwua* that is estimated based on the renewability rate of each geographical water cycle. The aggregated consumptive water use level for food production (water footprint inventory; WI) was found to be 4344 km^3/year, and the calculated global total WSF was 18,031 km^3 H_2Oeq/year, when considering the difference in water sources. According to the *fwua* concept, which is based on the land area required to obtain a unit volume of water from each source, the calculated annual impact can also be represented as 98.5×10^6 km^2. This value implies that current agricultural activities requires a land area that is over six times larger than global total cropland. We also present the net import of the WI and WSF, highlighting the importance of quantitative assessments for utilizing global water resources to achieve sustainable water use globally.

Keywords: agriculture; freshwater availability; life cycle impact assessment; virtual water trade; water footprint

1. INTRODUCTION

Water is a vital resource for human health, personal livelihoods, and quality of life. As recent water demand, especially in the agricultural sector, faces numerous challenges related to freshwater availability [1,2,3], quantifying the impacts of agricultural water use on freshwater availability is critical to improving sustainable water management. In regards to freshwater sustainability, the terrestrial renewable fresh water supply ($RFWS_{land}$) component has been proposed in the literature, and the human appropriation of accessible $RFWS_{land}$ has been estimated at 30% [4]. The planetary boundary concept states that limitations on global freshwater use range between 1100 and 4500 km^3/year [5,6]. These methods are based on the simple aggregated values of freshwater resources and their use; however, renewable freshwater resources exhibit spatiotemporal variability [7,8], implying that the potential environmental impacts of water use on freshwater availability also exhibit spatiotemporal variability. This finding shows that freshwater sustainability should be assessed in consideration of geographical distributions.

On the other hand, numerous studies have been conducted on the virtual flow of water through the international trade of food and other commodities. The virtual water concept has been developed to explain the virtual use of water through trading [9], and is has been adopted to estimate water dependency levels in countries using net virtual water imports [10]. Since the concept's establishment, water savings through international trade have been estimated globally [11,12,13]. Recently, global virtual water transfer has been associated with the trade of agricultural and industrial products [14,15]. Virtual water flow volumes can be used to estimate the global balance between water supply and demand and the volume of water that is compensated through international trade. However, the virtual flow of water should also be assessed in consideration of the geographical distribution of water resources for the sustainable trade of virtual water.

Some key technical terms and an influential proposal on water footprint assessment (WFA) and life cycle assessment (LCA) related to water have been introduced [16,17]. The International Organization for Standardization (ISO) defines the "water footprint" as a metric that quantifies potential impacts related to water. According to the ISO 14046, a summation of water use is referred to as a "water footprint inventory (WI)" [18]. Several methods have been proposed to estimate the potential environmental impacts of changes in water quantities. The groundwater footprint (GF) has been presented as an estimate based on the potential impacts of groundwater use in major groundwater basins, which is represented by the area needed to sustain groundwater use and groundwater-dependent ecosystem services [19]. The water stress index (WSI) is derived from the ratio of water withdrawals to freshwater availability and from a variation factor weighted by the spatial precipitation distribution [20]. While these methods consider geographical distributions of water resources, different water sources are not addressed (*i.e.*, only groundwater use for the GF and the aggregation of surface water and groundwater for the WSI). As each water source is uniquely renewable and available, the potential impacts of water use also vary by water source. The water unavailability factor (*fwua*) has been proposed as a means of characterizing water consumption based on its potential impacts, reflecting differences in water renewability levels according to place and water source [21]. Yano *et al.*[21] also conducted a case study to describe global food trade patterns via LCIA. However, the study was not based on the latest statistics or on multiple forcing datasets, and it did not involve a detailed discussion of country-based assessments of food production. In this study, we quantify the potential impacts of global food production on freshwater availability based on a detailed grid and country based evaluation. We assess the impact of international food trade from a LCIA perspective on sustainable water consumption, and we use the *fwua* concept, new trade statistics, and several forcing datasets to improve the robustness of the results.

2. MATERIALS AND METHODS

2.1. Calculation of the Water Footprint Inventory

In this study, the WI of food production is defined as the total evapotranspiration from croplands during the cropping period of each agricultural product. Water use for fertilizers, pesticides, and machine production is not considered. To estimate the WI for agricultural and livestock production, the H08 model, a global water resources model, was used [22]. As this model addresses the effects of human activities (e.g., reservoir operation and cropland irrigation), while considering different types of water sources, the geographical distributions of water resources and differences between water sources can be counted for in the calculation of WI. Water and heat balance calculations made at a daily interval reflect the temporal aspect if the renewability rate of water resources and water usage from each source. Meteorological forcing data, including air temperature, surface pressure, rainfall rate, snow fall rate, wind speed, radiation, and specific humidity level, are adopted as climatic input data at daily intervals. River basins are expressed based on the elevation and flow direction. Cropland and irrigated area are considered based on the global dataset [23,24]. The WI values of 58 major trade commodities in consideration of different water sources (*i.e.*, precipitation, stream flow, medium-sized reservoirs, and non-renewable and non-local blue water (NNBW)) are calculated at a resolution of 0.5° × 0.5°. NNBW is defined as a conceptual water source, and the WI of NNBW is defined as a component of water withdrawal that cannot be explained within the system boundaries. NNBW can be described as components of deep groundwater, lakes, glaciers, water diversions, and desalinated water that are not modelled in the H08 [12]. The NNBW value is known to be comparable to withdrawals reported for several aquifers in arid regions in the United States [12]. In this study, irrigated water from stream flows and medium-sized reservoirs are defined as the WI for surface water, and the irrigated NNBW value is defined as WI for groundwater.

WI flows were estimated for 58 major trade commodities derived from five major crops (*i.e.*, barley, maize, rice, soy, and wheat) and three livestock products (*i.e.*, beef, chicken, and pork) using FAOSTAT trade matrix data [25] (Table S1). The simulation settings used to calculate WI and WI flows were identical to those used in a previous study [10], although the meteorological forcing data and calculation period used differed. WATCH forcing data (WFD) [26], the third Global Soil Wetness Project (GSWP3) dataset [27], and Princeton's Global Meteorological Forcing Dataset (PFD) [28] were used as input data for calculating the average WI for croplands between 1991 and 2000. As the WFD covers the period running to 2000, the GSWP3 and PFD were used to calculate the WI for between 2001 and 2008. The average WI for each agricultural product between 1991 and 2000 was used to determine WI flows between 22 regions.

2.2. Water Scarcity Footprint Assessment

The calculated WI was converted into a WSF by multiplying the value by a characterization factor. As the *fwua* was developed to characterize the WI in consideration of different water sources, we used the *fwua* as the characterization factor for this study [21]. The *fwua* was devised by Yano *et al.* [21]. For the readers' convenience, the factor is described in brief below. See Yano *et al.* [21] for a detailed account and for the full formulation. The WSF of food production was determined as follows:

$$WSF = \sum_{x}\sum_{l}(fwua_{x,l} \times WI_{x,l}) \quad (1)$$

$$fwua_{x,l} = \frac{A_{x,l}}{A_{ref}} \quad (2)$$

where *x* denotes the water source (including precipitation, surface water, and groundwater); *fwua* is based on the land area required to obtain the reference volume of water from each water source; $A_{x,l}$ is the required land area per unit of time to obtain the reference volume for water from water source *x* at location *l*; and A_{ref} is the required land area per unit of time to obtain the reference volume of water from the reference condition [21]. In this study, the reference volume is defined as 1 m^3 of rainfall per year over an area of 1.0 m^2 (e.g., 1000 mm/year) based on the global mean annual precipitation level. The calculated WSF has global mean precipitation equivalent units (e.g., m^3 H_2Oeq). The *fwua* value for precipitation was calculated from the average annual precipitation for each domain using WFD, GSWP3, and PFD. The *fwua* values for surface water and groundwater were calculated from the average total runoff and sub-surface runoff levels of each domain using WFD, GSWP3, and PFD according to H08. All factors have a spatial resolution of 0.5° × 0.5°. The calculation period was identical to that used to determine the WI. The *fwua* values adopted herein do not account for the effects of water inputs from upstream regions, thus allowing the WSF to be assessed based on local water consumption levels. To prevent extremely high *fwua* values from increasing the total WSF, upper limits were defined at the 99th percentiles of each *fwua*. The upper *fwua* values for rainfed cropland precipitation, irrigated cropland precipitation, surface water, and groundwater were 5.0, 5.0, 100, and 200, respectively, for the study period.

3. RESULTS

3.1. Water Footprint Inventory and the Water Scarcity Footprint of Global Food Production

The global total WI related to global agricultural activities is shown in Figure 1 and in Tables S2 to S3. The total globally averaged WI reached 4344 km^3/year between 1991 and 2000, and 4592 km^3/year between 2001 and 2008. According to the t-test results that assume that two average values are equal in a null hypothesis, the p- value of the average values for the two periods was 0.6; therefore, differences between the WI results are not statistically significant. According to the individual water sources, the WI of precipitation in rainfed croplands accounted for more than 50 % of the total WI. India, China, the United States, Pakistan, and Indonesia generated the highest WI values during this period, *i.e.*, 790, 571, 484, 252, and 244 km^3/year, respectively. These five countries accounted for approximately 50% of the total global WI.

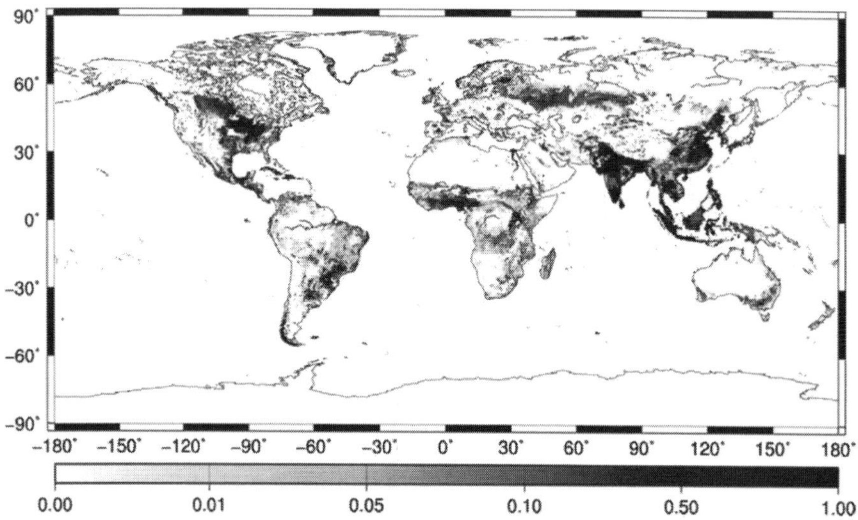

Figure 1. Global distribution of the annual water footprint inventory related to food production between 2001 and 2008 (km^3/year).

The global total WSF related to agricultural activities reached 18,031 km^3 H_2Oeq/year between 1991 and 2000, and 21,673 km^3 H_2Oeq/year between 2001 and 2008 (Tables S4 and S5). The WSF results for all water sources were greater than the WI values for these periods. Between 1991 and 2000, the WSFs for rainfed cropland precipitation, irrigated cropland precipitation, surface water, and groundwater were 1.2, 1.2, 7.7, and 21.2 times greater than the WI values, respectively. Although the WI for groundwater constituted only 8.5 % of the total WI, the WSF for groundwater accounted for 43.4 % of the total WSF. The highest WSF value was registered at 24.25° N, 33.25° E in Egypt. The WSF for groundwater reached 42.6 km^3 H_2Oeq/year, accounting for 95% of the total WSF of this domain. The *fwua* value for groundwater in this domain reached 171. This grid belongs to the Nubian Sandstone aquifer, which was reported as a non-renewable aquifer system [29]. Central America, India, Pakistan, Northern China, and Southeastern Australia registered high WSFs. Figure 2 shows the annual WSF and the WSF/WI ratio for each country between 1991 and 2000; the image was visually enhanced using a cartogram [30]. The area of each country is proportionate to the WSF map, and boundary lines have been maintained. Countries dependent on minimal renewable water resources are represented as large countries in this figure. China, India, Pakistan, the United States, and Egypt registered the highest WSFs, *i.e.*, 3415, 2061, 2408, 922, and 892 km^3 H_2Oeq/year, respectively. Saudi Arabia, Libya, Yemen, and Egypt registered the highest WSF/WI ratios, *i.e.*, 44.5, 38.2, 25.8, and 25.2, respectively. The Middle East and North African (MENA) countries tended to generate high WSF/WI ratios. Outside of the MENA region, Suriname and Chile registered WSF/WI ratios exceeding a value of 10. By contrast, Southeast Asia, Japan, Central Africa, and Brazil registered small areas with WSF/WI ratios of less than 1.0. The ratio of the WI for groundwater to the total WI for Brazil was founded to be less than 0.1%, whereas a value of 34.5% was found for Egypt.

The WSF results divided based on cropland areas within each country and basin (mm H_2Oeq/year) are shown in Figures S3 and S4. While the country-based results can be easily applied for assessment purposes using country-based statistical data, the basin-based results do not require one to consider the effects of water inputs from upstream areas of international river basins. The total WSF value for the United States reached 507 mm H_2Oeq/year (including 553 mm H_2Oeq/year for the Mississippi River, the largest basin in the country). The WSF of the Colorado River reached 4261 mm H_2Oeq/year, exceeding the national average value. As the cropland area of the river basin accounted for only 1.1% of the total cropland area for the US, Colorado River basin had a minor effect. For the countries with large cropland areas, the WSF values of local basins were more suited to sustainability assessments of local water issues.

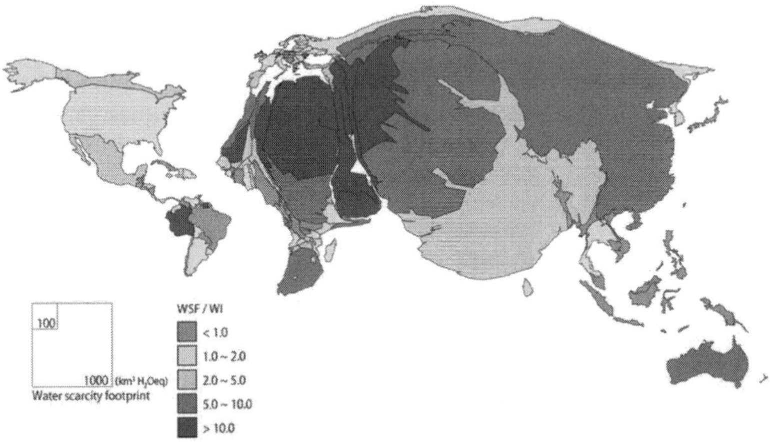

Figure 2. The water scarcity footprint for each country (area size) and the ratio of the water scarcity footprint to the water footprint inventory (colour coded).

3.2. Water Scarcity Footprint Flows Based on Food Trade

WSF flows from 2010 that were divided by country were aggregated into 22 regions based on a previous study [18].Figure 3 shows (a) net exports of the WSF for all water sources and (b) irrigated groundwater alone. The "total" value represents the total net flows between the 22 regions, and "other" signifies the total flows below threshold values, *i.e.*, 10 km^3 H_2Oeq/year for all water sources and 0.5 km^3 H_2O/year for irrigated groundwater. The numbers shown in parentheses are the standard deviations of the calculated values between the three meteorological forcing datasets. Global total net flows of WI between the 22 regions were calculated at 524.8 km^3/year. Previous studies have estimated global gross flows of WI related to crop trades as 695 km^3/year for 1995–1999 and global net exports as 355 to 545 km^3/year for 2000 [12,21,31]. Although there is not full agreement between the simulation settings (the period, net and gross values, countries, and crops examined in this study), it appears that WI flows were adequately estimated. The WSF flows tended to generate larger arrows than the WI for both types of water sources. The total WSF flows were 1.8 times greater than the total WI flows. WSF water source, precipitation, surface water, and groundwater values were 1.1, 8.2, and 13.8 times greater than the WI values, respectively. In regard to each region, WSF flows for Northern Africa, Eastern Asia, Southern Asia, Western Asia, North America, and Oceania were 15.5, 15.6, 4.9, 11.7, 1.8, and 3.4 times greater than the WI values, respectively. In comparison, WSF flows for the Caribbean, South America, and Northern Europe were 0.6, 0.9, and 0.8 times greater than the WI values, respectively, and in these regions, food production levels were dependent on the presence of abundant water resources. North America, South America, Oceania, and Southern Asia were characterized by net exports of 345.4, 187.8, 118.6, and 109.1 km^3 H_2Oeq/year, respectively, in 2010. Relative to previous case study results on WSF flows in 2000 [22], the imported value of the WSF for Eastern Asia has been increasing. The imported WSF from North America grew from 88.5 km^3 H_2Oeq/year in 2000 to 115.4 km^3 H_2Oeq/year in 2010 and that from South America increased from 12.6 km^3H_2Oeq/year in 2000 to 56.8 km^3 H_2Oeq/year in 2010. This fact is mainly attributed to the growth of the Chinese economy. According to the FAO, the net importing value of crops and livestock products of China increased from 6.5 billion USD in 2000 to 55.4 billion USD in 2010.

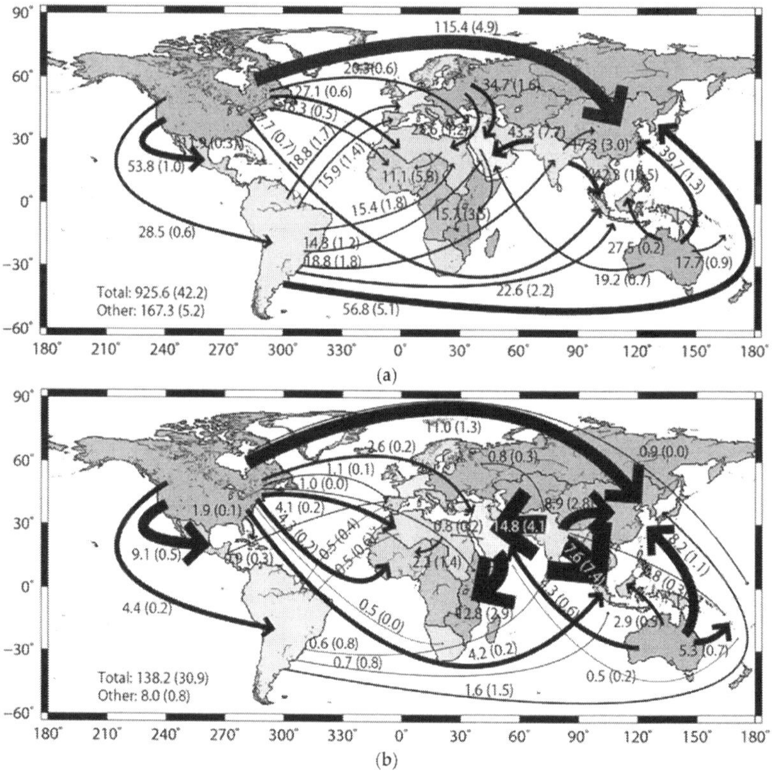

Figure 3. Water scarcity footprint flows due to international food trade between 22 regions for (**a**) all water sources (exceeding 10 km^3 H_2Oeq/year) and (**b**) groundwater (exceeding 0.5 km^3 H_2Oeq/year).

The WSF/WI ratios for groundwater for North Africa, North America, South America, Eastern Asia, and Southern Asia were recorded as 18.5, 9.1, 27.7, 14.5, and 14.0, respectively. The largest flows of the WSF from groundwater were reached 46.4 km^3 H_2O/year for North America, 43.8 km^3 H_2Oeq/year for southern Asia, 21.3 km^3 H_2Oeq/year for Oceania, and 14.5 km^3 H_2Oeq/year for Western Asia. South America, which generated the second largest WSF flow for all water sources, registered a net WSF export level of only 5.3 km^3 H_2Oeq/year for groundwater. This implies South America did not export food produced from scarce water resources, although large quantities of food produced from water sources with lower potential impacts were exported during the study period.

4. DISCUSSION

The estimated global total WSF presents valuable information for understanding the sustainability of existing agricultural systems. As the *fwua* value is based on the land area required to obtain the reference volume of water, the WSF/WI ratio contains information on the required land area to maintain current agricultural conditions in terms of water availability. The WSF/WI ratio can also be described as a characterization factor weighted by the volume of water use. The WSF/WI ratios for rainfed cropland precipitation, irrigated cropland precipitation, irrigated cropland surface water, and irrigated cropland groundwater were calculated as 1.2, 1.2, 7.7, and 21.2, respectively. Rainfed and irrigated cropland areas covered 12.4×10^6 km^2 and 2.8×10^6 km^2 [23,24], respectively. The 18,031 km^3 H_2Oeq/year value calculated for the WSF can also be referred to as 98.5×10^6 km^2 of the required land area by multiplying each cropland area by the corresponding WSF/WI ratio. This value was 6.6 times greater than the 14.9×10^6 km^2 value calculated for the total cropland area [23] and was 6.4 times greater than the 15.4×10^6 km^2 value calculated for the total arable and permanent cropland area of 2010 [32].

Figure 4 shows the net WI and WSF import levels due to international food trade (exports are negative), the gross domestic product (GDP) per capita [33], and the ratio of irrigated croplands to cumulative croplands for each region. Melanesia, Micronesia, and Polynesia were excluded from the analysis, as these regions generated low proportions. As noted above, South America's WSF value is smaller than its WI, implying that the region produced food using abundant water resources. In comparison, Oceania and North America generated larger WSFs relative to their WI values. Southern Asia was a net importing region in terms of its WI; however, it was a net exporting region in terms of its WSF. In other words, Southern Asia exports foods produced from scarcer water resources than the water resources used to produce imported foods. GDP per capita values serve as a justification for why Oceania and North America exported foods using scarce water resources. For example, a high GDP creates technological capacities to develop new water resources, to maintain high water-use efficiency levels, and to secure low water use costs due to the presence of competitive industry structures. However, Southern Asia exported food produced from scarce water resources due to the country's high ratio of irrigated cropland. Southern Asia's irrigated ratio reached 37%, exceeding values of 10% and 13% for Oceania and North America, respectively. This high ratio resulted in higher water-use efficiency levels, implying that the aggressive use of scarce water resources occurred in Southern Asia.

Figure 5 shows the renewable water resources [32] and WI per capita for each region. Although South America and Central Africa prevented abundant water resources per capita, the ratio of water utilization remained low. The sustainable use of water at a global scale requires the utilization of abundant water resources with lower potential impacts (especially in these regions) in consideration of supply, demand, and sustainable water use. This quantitative assessment of WI and WSF can contribute to strategies on how to utilize global water resources to achieve global sustainability. Domestic water resources may not be utilized adequately for the following reasons:

- Irrigated croplands are limited due to low GDP levels (e.g., South America and Central Africa), resulting in limited water resource utilization;
- Industrial structures have not been developed due to low GDP levels, which has led to minimal competitive and small-scale agricultural practices and low levels of water resource utilizations;
- Water resource development is difficult to achieve due to low GDP and low technological capacity levels, resulting in low water resource utilization;
- High costs of infrastructure use imply high costs for water use, resulting in low water resource utilization; and

Climatic conditions are not conductive to intensive agriculture (e.g., Northern Europe [34]), causing individuals to not utilize domestic water resources adequately.

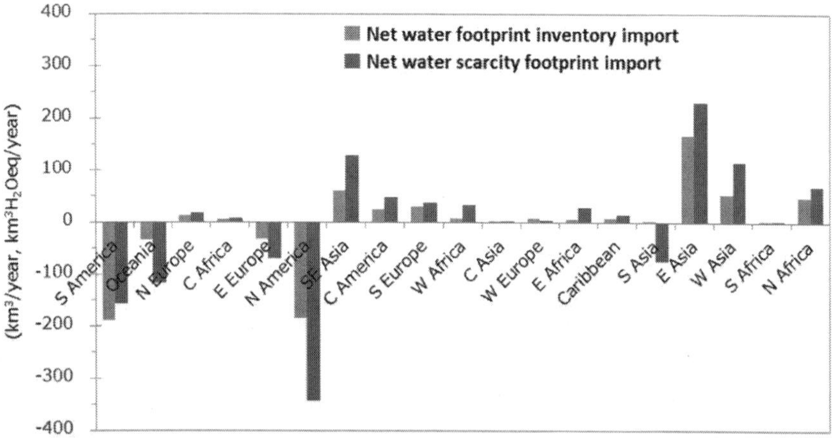

(a)

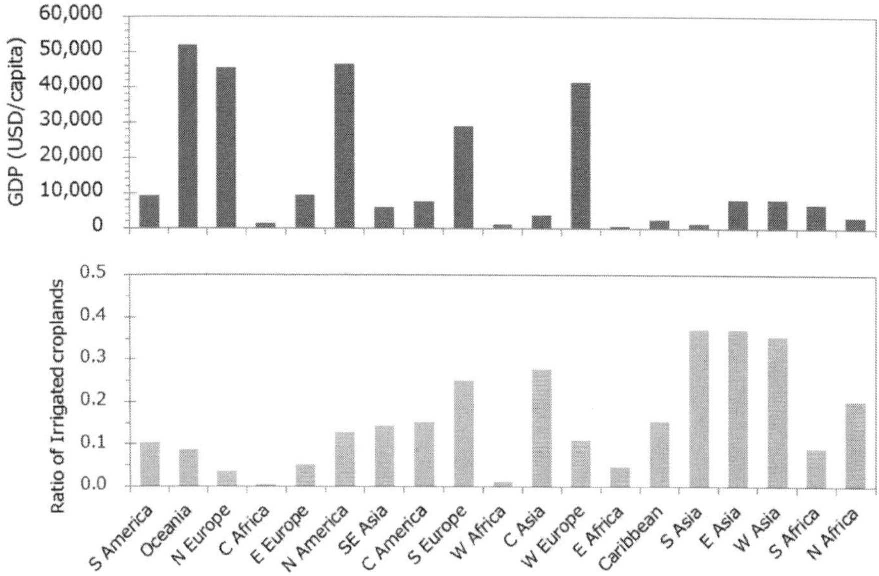

(b)

Figure 4. (**a**) The net import of the water footprint inventory and water scarcity footprint and (**b**) the GDP per capita for 2010 and the ratio of irrigated cropland to total cropland by region.

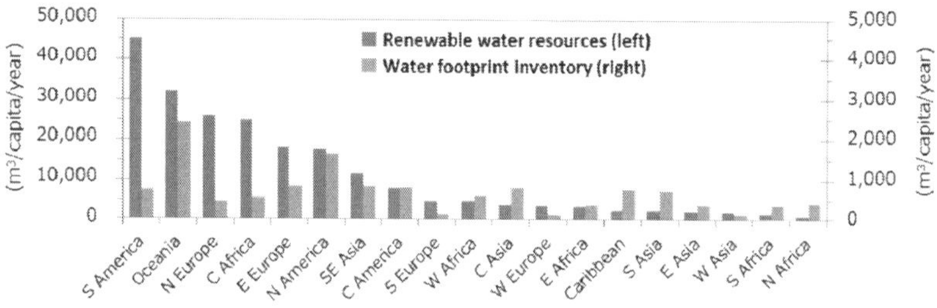

Figure 5. Renewable water resources and water footprint inventory per capita by region.

All of these phenomena are influenced by both social and natural factors. However, resolving uneven GDP distributions could effectively promote global water-use efficiency.

In regards to hydrological modelling, the state of current practices has been reviewed and the scientific and data-related challenges have been considered with a focus on human-water activities manifested through water resource management [35,36]. The H08 is used as one of the latest large-scale hydrological models to show how international trade helps control regional water demand. The results of our calculations include uncertainties due to modelling assumptions. In considering the uncertainties in meteorological forcing data, this study adopts an average result value using three meteorological forcing datasets. To evaluate uncertainties in groundwater withdrawal levels, we used the NNBW measure that has been validated as comparable to groundwater withdrawal levels reported for several arid aquifers. Although these results include uncertainties, this study presents new information and insights on water footprint and international food trade patterns from an LCA point of view. It is valuable to quantitatively assess the potential impacts of water use and virtual flows related to international food trade at a global scale in consideration of different water sources.

5. CONCLUSIONS

Potential impacts on global food production freshwater availability (WSF) and flows of WSF related to international food trade were estimated using the *fwua*, which reflects the difference in water renewability rates by source. The global WSF of food production reached 18,031 km^3 H_2Oeq/year and was also described as 98.5×10^6 km^2 of the required land area, which is equivalent to 4.2 times the value of the summation of the WI. Although groundwater constituted only 8.5% of the total WI, it accounted for 43.4% of the total WSF. China, India, Pakistan, the United States, and Egypt exhibited the highest WSFs, showing dependence on minimal renewable water resources. WSF flows via food trade are 1.8 times greater than the total WI flows. In total, 925.6 km^3 H_2Oeq of WSF are traded each year. The WSF imported by Eastern Asia is growing mainly because of the rapid growth of the Chinese economy. Southern Asia is a net WI importing region, however, it is also a net WSF exporting region, which implies its dependence on scarcer water resources for food production rather than on water used to produce imported foods. Sustainable water use requires the utilization of abundant water resources at a global scale, and uneven GDP distributions can be resolved to effectively promote global water-use efficiency. The assessment of WSF showed a new insight that net WI importing regions can be net WSF exporting regions considering the difference in water sources. The results illustrate the importance of quantitative assessments of WI and WSF for utilizing global water resources to achieve global sustainability with lower potential impacts.

ACKNOWLEDGMENTS

The authors thank the WATCH projects, Graham Weedon, Hyungjun Kim, the Terrestrial Hydrology Research Group, Justin Sheffield, and Ryo Inoue for providing datasets. This research was supported by the Japan Society for the Promotion of Science KAKENHI via by a Grant-in-Aid for Scientific Research (S) (23226012) and the Environment Research and Technology Development Fund (S-10, S-14) of the Ministry of the Environment, Japan.

AUTHOR CONTRIBUTIONS

Shinjiro Yano and Taikan Oki conceived of and designed the experiments; Shinjiro Yano performed the experiments, analysed the data, and wrote the paper; N.H. contributed materials and analysis tools; Norihiro Itsubo provided technical support and conceptual advisement on the life cycle assessment. All of the authors read and approved the final manuscript.

CONFLICTS OF INTEREST

The authors declare no conflict of interest.

REFERENCES

1. The 2030 Water Resources Group. *Charting our Water Future: Economic Frameworks to Inform Decision-Making*; McKinsey & Company: New York, NY, USA, 2009.
2. Foley, J.A.; Ramankutty, N.; Brauman, K.A.; Cassidy, E.S.; Gerber, J.S.; Johnston, M.; Mueller, N.D.; O'Connell, C.; Ray, D.K.; West, P.C.; *et al.* Solutions for a cultivated planet. *Nature* **2011**, *478*, 337–342.
3. Food and Agriculture Organization of the United Nations (FAO). *The State of the World's Land and Water Resources for Food and Agriculture (SOLAW)—Managing Systems at Risk*; FAO: Rome, Italy, 2011.
4. Postel, S.L.; Daily, G.C.; Ehrlich, P.R. Human appropriation of renewable fresh water. *Science* **1996**, *271*, 785–788.
5. Rockström, J.; Stefen, W.; Noone, K.; Persson, A.; Chapin, F.S.; Lambin, E.F.; Lenton, T.M.; Scheffer, M.; Folke, C.; Schellnhuber, H.J.; *et al.* A safe operating space for humanity. *Nature* **2009**, *461*, 472–475.
6. Gerten, D.; Hoff, H.; Rockström, J.; Jägermeyr, J.; Kummu, M.; Pastor, A.V. Towards a revised planetary boundary for consumptive freshwater use: Role of environmental flow requirements. *Curr. Opin. Environ. Sustain.* **2013**, *5*, 551–558.
7. The United Nations Educational, Scientific and Cultural Organization (UNESCO). *Managing Water under Uncertainty and Risk (Vol. 1)*; The United Nations World Water Development Report 4; UNESCO: Paris, France, 2012.
8. Hoekstra, A.Y.; Mekonnen, M.M.; Chapagain, A.K.; Mathews, R.E.; Richter, B.D. Global monthly water scarcity: Blue water footprints *versus* blue water availability. *PLoS One* **2012**, *7*, e32688.
9. Allan, J.A. Virtual water: A strategic resource global solutions to regional deficits. *Groundwater* **1998**, *36*, 545–546.

10. Hoekstra, A.Y.; Hung, P.Q. *Virtual Water Trade, A Quantification of Virtual Water Flows between Nations in Relation to International Crop Trade*; Value of Water Research Report Series No. 11; UNESCO-IHE: Delft, The Netherland, 2002.

11. Aldaya, M.M.; Llamas, M.R. *Water Footprint Analysis for the Guadiana River Basin*; Value of Water Research Report Series No. 35; UNESCO-IHE: Delft, The Netherland, 2008.

12. Hanasaki, N.; Inuzuka, T.; Kanae, S.; Oki, T. An estimation of global virtual water flow and sources of water withdrawal for major crops and livestock products using a global hydrological model. *J. Hydrol.* **2010**, *384*, 232–244.

13. Hoekstra, A.Y.; Mekonnen, M.M. The water footprint of humanity. *PNAS* **2012**, *109*, 3232–3237.

14. Oki, T.; Kanae, S. Global hydrological cycles and world water resources. *Science* **2006**, *313*, 1068–1072.

15. Mekonnen, M.M.; Hoekstra, A.Y. *National Water Footprint Accounts: the Green, Blue and Grey Water Footprint of Production and Consumption*; Value of Water Research Report Series No. 50; UNESCO-IHE: Delft, The Netherland, 2011.

16. Boulay, A.M.; Hoekstra, A.Y.; Vionnet, S. Complementarities of water-focused life cycle assessment and water footprint assessment. *Environ. Sci. Technol.* **2013**, *47*, 11926–11927.

17. Manzardo, A.; Mazzi, A.; Loss, A.; Butler, M.; Williamson, A.; Scipioni, A. Lessons learned from the application of different water footprint approaches to compare different food packaging alternatives. *J. Clean. Prod.* **2016**, *112*, 4657–4666.

18. International Organization for Standardization (ISO). *ISO 14046 Environmental Management, Water Footprint—Principles, Requirements and Guidelines*; ISO: Geneva, Switzerland, 2014.

19. Gleeson, T.; Wada, Y.; Bierkens, M.F.P.; van Beek, L.P.H. Water balance of global aquifers revealed by groundwater footprint. *Nature* **2012**, *488*, 197–200.

20. Pfister, S.; Koehler, A.; Hellweg, S. Assessing the environmental impacts of freshwater consumption in LCA. *Environ. Sci. Technol.* **2009**, *43*, 4098–4104.

21. Yano, S.; Hanasaki, N.; Itsubo, N.; Oki, T. Water Scarcity Footprints by Considering the Differences in Water Sources.*Sustainability* **2015**, *7*, 9753–9772.

22. Hanasaki, N.; Kanae, S.; Oki, T.; Masuda, K.; Motoya, K.; Shirakawa, N.; Shen, Y.; Tanaka, K. An integrated model for the assessment of global water resources—Part 1: Model description and input meteorological forcing. *Hydrol. Earth Sys. Sci.* **2008**, *12*, 1007–1025.

23. Ramankutty, N.; Evan, A.; Monfreda, C.; Foley, J.A. Farming the planet: 1. Geographic distribution of global agricultural lands in the year 2000. *Glob. Biogeochem. Cycles* **2008**, *22*.

24. Siebert, S.; Doll, P.; Hoogeveen, J.; Faures, J.M.; Frenken, K.; Feick, S. Development and validation of the global map of irrigation areas. *Hydrol. Earth Sys. Sci.* **2005**, *9*, 535–547.

25. The Food and Agriculture Organization Corporate Statistical Database (FAOSTAT). Available online: http://faostat.fao.org/ (accessed on 5 August 2014).

26. Weedon, G.P.; Gomes, S.; Viterbo, P.; Shuttleworth, W.J.; Blyth, E.; Österle, H.; Adam, J.C.; Bellouin, N.; Boucher, O.; Best, M. Creation of the WATCH forcing data and its use to assess global and regional reference crop evaporation over land during the twentieth century. *J. Hydrometeorol.* **2011**, *12*, 823–848.

27. Kim, H.; Yoshimura, K.; Chang, E.; Famiglietti, J.S.; Oki, T. Century long observation constrained global dynamic downscaling and hydrologic implication. In Proceedings of the AGU Fall Meeting 2012, San Francisco, CA, USA, 3–7 December 2012; American Geophysical Union: Washington DC, USA, 2012; GC31D-02.

28. Sheffield, J.; Goteti, G.; Wood, E.F. Development of a 50-yr high-resolution global dataset of meteorological forcings for land surface modeling. *J. Clim.* **2006**, *19*, 3088–3111.

29. Arabi, N.E. Environmental Management of Groundwater in Egypt via Artificial Recharge Extending the Practice to Soil Aquifer Treatment (SAT). *Int. J. Environ. Sustain.* **2012**, *1*, 66–82.

30. Inoue, R.; Shimizu, E. A new algorithm for continuous area cartogram construction with triangulation of regions and restriction on bearing changes of edges. *Cartogr. Geogr. Inf. Sci.* **2006**, *33*, 115–125.

31. Chapagain, A.K.; Hoekstra, A.Y. *Virtual Water Flows between Nations in Relation to Trade in Livestock and Livestock Products*; Value of Water Research Report Series No. 13; UNESCO-IHE: Delft, The Netherland, 2003.

32. AQUASTAT. Available online: www.fao.org/nr/aquastat/ (accessed on 23 July 2014).

33. World Economic Outlook Databases. Available online: http://www.imf.org/external/ns/cs.aspx?id=28 (accessed on 1 February 2014).

34. Maracchi, G.; Sirotenko, O.; Bindi, M. Impacts of Present and Future Climate Variability on Agriculture and Forestry in the Temperate Regions: Europe. *Clim. Chang.* **2005**, *70*, 117–135.

35. Nazemi, A.; Wheater, H.S. On inclusion of water resource management in Earth system models—Part 1: Problem definition and representation of water demand. *Hydrol. Earth Syst. Sci.* **2015**, *19*, 33–61.

36. Nazemi, A.; Wheater, H.S. On inclusion of water resource management in Earth system models—Part 2: Representation of water supply and allocation and opportunities for improved modeling. *Hydrol. Earth Syst. Sci.* **2015**, *19*, 63–90.

2

CHAPTER

Heavy Metals in Agricultural Soils of Central Ethiopia: The Contribution of Land Use Types and Organic Sources to Their Variability

Nigatu Alemayehu Minase[1*], Mary M. Masafu[2], Abule Ebro Geda[1], Azage Tegegne Wolde[1]

[1]*International Livestock Research Institute, Addis Ababa, Ethiopia*

[2]*University of South Africa, Pretoria, South Africa*

ABSTRACT

In the present study, about 94% of the arable land had a phosphorus deficiency (available P < 10 mg·kg^{-1}). To compensate for this deficiency, high amounts of phosphorous fertilization were applied to soils to get better crop yield. In the past two decades, soil organic matter has sharply declined from 34.8 g/kg in late 1980s' to 12.0 g/kg in 2010. The soil has moderate to high cation exchanging capacity (CEC) in the range of 45 - 58 Meq/100 gm. The low soil organic carbon, high CEC and high phosphorus fertilization could lead to poor heavy metal availability to plants. Nevertheless, DTPA extractable concentrations were 40.5 mg/kg for copper, 35.5 mg/kg for iron, 134 mg/kg for manganese, 2.1 mg/kg for nickel, 2.6 mg/kg for lead and 53.5 mg/kg for zinc. All the available micronutrients seem to be adequate for plant growth. In terms of toxicity, the values of Cu, Fe, Mn, Zn, Ni and Pb found in the present investigation were lower than those of above critical value set by the international guidelines and other researchers. Variations in heavy metal content in the soil were attributed due to differences in land use types and disparity in organic sources amongst the plant and animal composites. For better crop productivity, more organic matter should be added to the soils. The traditional nutrient cycling practices, such as manure application, allow crop residue to decay on fields, fallowing and rotational cropping has to be followed to rehabilitate the soil and buildup both macro-and micro-nutrients to their natural level.

Keywords: Phosphorous, Toxicity, Micronutrients, Organic Carbon, Nutrient Cycle

1. INTRODUCTION

The central Ethiopian highland is coming under increased pressure for the expansion of factories, export abattoirs, flower farms and other processing plants. These factories produce packed and semi-processed foods, tanneries and finished products like textiles, cosmetics, detergents, cement, metal and beverage products and largely congregate in 100 km radius of Addis Ababa, the capital of Ethiopia. This has resulted in shrinkage of the adjoining smallholders' farms that used for the production of food in the land they have used for crop cultivation and cattle grazing. The result is that at the present rate of expansion due to both urbanization and industrialization, the survival of these farmers and their farm land is at stake. Most of the cultivated area is put to food crops like teff (Eragrostis abyssinica) and wheat (Triticum aestivum) and marsh and hillsides for sheep, goats and cattle grazing. The central Ethiopian highland is also a potential cereal producing zone and it is the highest supplier of high quality crop like teff to Addis Ababa and other towns in the country. Ethiopia also needs to boost food production and this is believed to be achieved by increased crop production per unit area to meet the needs of the increasing population. In the future, increased and sustained food production will need more fertilizer amendments. Under heavy or continuous fertilization, micronutrients and other heavy metals could also be introduced into the soil [1] .

In Ethiopia context, researches on heavy metal toxicity and availability to crops have been limited around few towns' along the river banks that are contaminated by factory effluents and have been used for vegetable production [2] - [7] . There has been a growing concern about food quality and safety of agricultural products obtained from such contaminated sites.

It is clear that soils under repeated contamination from different pollutants eventually lose their resilience and are unable to provide good quality food products [1] , but there is also a risk in productivity and quality when the level of heavy metals are below crop requirement. Though, risk of contamination from heavy metals in rural Ethiopia was minimal [1] [8] , acquisition of base line data on their concentrations would be useful in future in decision making about the type of organic and/or inorganic fertilizers that could be applied [1] .

Given the important role of micronutrients in plant growth and animal nutrition and the deleterious effects of heavy metal contamination in the environment, the determination of background levels of these elements is therefore an important guide to prevent physiological disorders in grazing animals as well as serving as initial threshold levels for studying micronutrient concentrations and heavy metal pollution in future [1] . The available micronutrients and heavy metals are of more importance since they constitute the nutrition pool taken up by the plants.

The naturally occurring background levels of heavy metals in uncontaminated soils usually occur in trace amounts related to the biogeochemistry of the parent materials [9] . Heavy metal content and distribution in soils are influenced by several factors including organic matter content, types of land management, particle-size distribution, parent material, drainage, soil age, vegetation and aerosol input [10] [11] . In the present study we would like to streamline the type of specific organic sources and land use types that could have influenced the content and distribution of heavy metals.

In the current study six heavy metals (Cu, Fe, Mn, Ni, Pb and Zn) were analyzed for their level of heavy metal concentration in smallholders' farms in Central Ethiopia. Fe and Cu are very important and mobile elements which have been widely studied globally, Ni and Zn are mobile in the soil are essential for plant growth, while Pb (although a relatively immobile metal) is considered a very serious environmental contaminant [12] . In the light of all the above, soil samples were collected and analyzed to emphasis the impact of applying different organic sources in to soil and explain the role of land management in heavy metal load in the soil. The spilling of industrial effluent into neighborhood crop farms through water flows and flooding were highlighted to indicate the future shadow of industrial development in rural environment. It will be obvious that with increasing urbanization and industrialization in Central Ethiopian highland the future environmental issue will be on land utilization of sludge and municipal waste management. Therefore, determination of heavy metal concentration is an important guide to protect the public health hazards as well as serve as initial threshold levels for studying heavy metal pollution in the future.

The specific purpose of this study was to:

Ÿ Determine the heavy metal concentration in agricultural soils of the study area.

Ÿ Compute the variation of heavy metal in different land use at the soil depths of 0 to 15 cm and 15 to 30 cm.

Ÿ Compare heavy metal variation under the influence of different organic sources at the soil depths of 0 to 15 cm and 15 to 30 cm.

2. MATERIALS AND METHODS

2.1. Location of Sites

Ada district lies between longitudes 38°51' to 39°04'E and latitudes 8°46' to 8°59'N covering an area of 1750 km^2 east of Addis Ababa. Most of the land (90%) was plain highlands ranging between 1600 to 2000 meters above sea level. The district is characterized by sub-tropical climate and receives 860 mm rain per annum [13] . Mean annual temperatures ranged from about 8°C to 28°C. Black clay vertisol soil was the dominant soil type. Ada district is became an industrial hot spot due to its proximity to Addis Ababa and situated on the main road that connects Ethiopia to sea port of Djibouti as well as serve as a trade corridor to agricultural commodity market between Central, Southern and Eastern Ethiopia (see Figure 1).

2.2. Soil Sample Categories

In the study area 15 soil samples sites were taken at the depth of 0 to 15 cm and 15 to 30 cm. Four subsamples were taken from each sample site. Subsamples of each site were mixed and pooled in one bag. The sample sites were grouped in two categories by [14] according to (a) land use and management (b) sources of organic matter (see Table 1).

The sites were also categorized according to their land use as follows (a) Sites with chemical disposal effluents reach farms through water flow and folds from factory outlets such as detergents, steel, plastic bags, and food processing plants, Number of sites or N = 4; (b) Grazing land these are communal grazing land allocated for livestock grazing only, N = 2; (c) Croplands that were solely used for crop cultivation for more than 3 decades, N = 7; (d) Undisturbed land that was free of human and animal interference for more than 3 decades, such as church and research compounds and areas enclosed for conservation and biodiversity, N = 2.

The other grouping of the sample sites was the three classes of organic sources such as: (a) Animal debris from slaughter house waste disposal places sites, N = 5; (b) Plant debris which included decayed crop residues and compost from crop rotation and non-organic sources, N = 4; (c) no external plant and animal organic additions, N = 6. The last category was helpful in the analysis of heavy metal loads in soils and their mobility due to natural sediments and soil contents.

2.3. Sampling and Chemical Analysis

Soil samples were collected purposely to describe the different land management practices and different organic sources such as degradation, area enclosure, crop history, tillage type, and levels of fertilization, residues left in the field and presence or absences of compost or manure applications. Samples were taken at the depths of 0 to 15 cm and at 15 to 30 cm with four subsamples at each composite site. Thus, a composite sample was made from four pooled subsamples. The four subsamples at each site were pooled for homogeneity and stored in a plastic bag and transported to Debrezeit Research Centre for soil chemical analysis in the Soil Laboratory within two hours of collection.

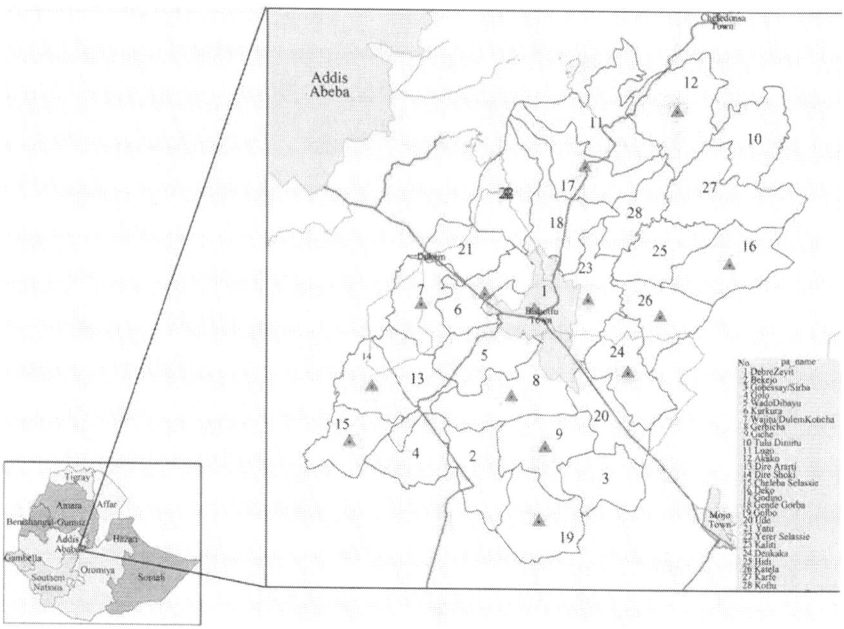

Figure 1. Location of the study area and sample sites denoted with a symbol of green triangle.

Table 1. Sample sites and their characteristics.

Case number	Sample site	Land use	Source of organic manure
1	DZ research center	Uncultivated	None
2	Private commercial Farm	Cropland	Animal
3	Crop rotation plot	Cropland	Plant
4	Minimum tillage	Cropland	Plant
5	Farmyard manure plot	Cropland	Animal
6	Compost plot	Cropland	Plant
7	Recycling residue	Cropland	Plant
8	Communal grazing	Grazing	Animal
9	Fenced	Uncultivated	None
10	Mechanized farming	Cropland	None
11	Effluent contaminated	Chemical disposal	None
12	Effluent contaminated	Chemical disposal	None
13	Effluent contaminated	Chemical disposal	None
14	Abattoir waste disposal	Chemical disposal	Animal
15	Swampy disposal site	Grazing	Animal

The pooled samples were air dried at ambient temperature and crushed to pass through a 2 mm stainless sieve. Then portions of pooled samples (about 50 cm) were ground to pass through a 0.149 mm sieve and stored in plastic bags prior to the chemical analysis. The pooled samples were analyzed for the following properties: pH, CEC, soil organic matter (SOM), available P, Fe, Mn, Ni, Cu, Zn and Pb. All the micro-nutrients were determined by DTPA extract through Atomic Absorption Spectrometer [15] . Soil pH was determined by a pH meter with a soil/water ratio of 1: 2.5. Available P was extracted by shaking 1.5 gm of soil sample for 30 minutes at 20°C in 100 mL of a 42% $NaHCO_3$ solution at pH 8.5 according to [16] . The organic content in the soil was determined based on the method from [17] . CEC was determined using ammonia distillation [18] . Exchangeable bases such as Ca were measured using 1N ammonium acetate per unit of meq/100 gm of soil.

2.4. Statistical Analysis

Statistical analysis and graphs were performed using [19] . General linear model (GLM) was used to compare the mean difference of different treatments. Correlation (r) and analysis of variance (ANOVA) were also calculated. The univariate analysis of variance was used to determine the mean, standard deviation and significant differences of heavy metal concentrations in the soil.

Cluster analysis was used in the grouping of different sample sites with various levels of heavy metal content in the soil. Cluster analysis or clustering is the task of assigning a set of objects into groups (clusters) so that the objects in the same cluster are more similar to each other than to those in other clusters [20] . At different distances, different clusters form a dendrogram or hierarchical clustering [21] . In the current study euclidean distance, the most commonly chosen type of distance was used [20] . It is preferred for the geometric distance in the multidimensional space. It is computed as:

$$\text{distance}(x,y)=\left\{\sum_i (X_i - Y_i)^2\right\}^{1/2}.$$

In a dendrogram, the X-axis marks the distance at which the clusters merge, while the objects are placed along the Y-axis so that the clusters don't mix. Popular choices are known as single-linkage clustering that use minimum distance between objects [20] .

3. RESULTS AND DISCUSSIONS

The concentrations of heavy metals in the soil and their impacts on ecosystems can be influenced by many factors such as parent material, climate and anthropogenic activities ([12] . Heavy metals may be added to soils in chemical fertilizers and pesticides, soil amendments (e.g. Lime and gypsum) or organic fertilizers [22] . Cluster analysis for heavy metals assists in tracing the origins of elevated levels of heavy metals in soil. This paper explains the origin and sources of heavy metals in the study area, the contribution of different organic sources to heavy metal loads, and describes the relationship of heavy metals to land use type.

3.1. Origins of Trace Elements in Soils of the Study Area

The important consideration in the heavy metal identification in soils was to determine the original sources such as natural weathering, industrial processing, use of metal components in commercial processes, aerial deposition of smelting materials, leaching from garbage and solid waste dumps, the application of animal products, or some other sources [23] .

Heavy metals in smallholder agriculture were mostly related to the addition of organic and inorganic nutrients to the soil. Most of the materials were added to soil for agricultural purposes such as manure, lime and other inorganic fertilizers have low trace element levels and when they were applied at normal rates and they did not affect overall concentrations of trace elements in soil. However, when they were applied at a higher rate they raised the concentration of trace elements in the soil. Other sources of inputs that increased the concentration of heavy metals in soil, included herbicides, fungicides, irrigation water, biosolid application, dredged materials, fly ash and municipal composts [24] .

In the current study, the cluster analysis sample sites were classified into seven groups, after assigning average weight based on Eculdian distance measurement for all variables [20] . It was found that major sources of heavy metals were ordered in their importance as follows in sample sites 13 - 15, 9 and 3. Site 13 represents steal rolling effluent discharge, site 14 was a place for municipal abattoir disposal where animal waste was disposed, site 15 located in swampy area where waste and dead animals' bodies were disposed, site 9 is a location fenced off from animal and human entrance and site 3 was a farm plot where crop rotation was practiced for a long period by rotating pulses with cereals. The result of the cluster analysis showed that the three major factors that contributed to heavy metal stock in soils in the study area (contributed for more than 60% variation) were anthropogenic factors such as the flow of factory effluents to farms, addition of organic products and natural factors such as soil sediments as shown in Figure 2.

The physical and chemical characteristics of heavy metal concentrations in topsoil of the study area are shown in Table 2. The pH of the topsoil was between 6.38 and 8.08 (average 7.00) at both soil depths. The soils had a normal pH but about 56% of samples had pH > 7.0. Soil scientists have reported that the pH was an essential factor that influenced the cation mobility and regulated the solubility of heavy metals in the soil as most of the metals tend to be available to plants in acidic pH. [25] . Higher soil pH is not favorable for the transfer of heavy metals from the soil to plants [26] . From Table 2, it is possible to argue that at both soil depths heavy metals in the top soil are available to plants due to normality of the pH. CEC refers to the preservation and supply of soil fertilizer and buffering capacity. The mean of the soil CEC in current study was 46 ± 12.4 mmol·kg^{-1}. The range was between 45.00 and 58.80 mmol·kg^{-1}, implying that the soils had higher CEC [27] with moderate capacity to retain and maintain soil fertility. Soils SOM (Soil Organic Matter) was 12.00 g/kg when this value compared with the findings reported by [28] of 34.8 g/kg in the identical site, SOM has sharply decline in the past two decades. For all soils pH, available P, Ca and CEC generally remained

```
                     Rescaled Distance Cluster Combine

 C A S E     0         5        10        15        20        25
 Label  Num  +---------+---------+---------+---------+---------+

          6   -+
         11   -+-+
          2   -+ |
         10   -+ +-+
          8   ---+ |
          1   ---+ +-----------+
          5   -+   |           |
         12   -+---+           +-----------+
          4   -+   |           |           |
          7   -----+           |           +-------------------+
          9   ---------------+-+           |                   |
         14   ---------------+             |                   |
         13   -----------------------------+                   |
          3   ---------------------------------------+---------+
         15   ---------------------------------------+
```

Figure 2. A dendrogram showing hierarchical cluster analysis.

Table 2. Aggregated statistical value of important soil variables.

Variables with soil depth		Mean	SD	Minimum	Maximum
Soil carbon %	0 - 15 cm	1.40	1.03	0.07	3.90
	15 - 30 cm	0.99	0.73	0.07	2.48
Cation exchange	0 - 15 cm	44.95	14.22	14.80	58.80
Meq/100gm	15 - 30 cm	46.88	10.58	27.80	58.80
Electrical cond.	0 - 15 cm	93.62	100.12	19.23	435.00
Mic. Siem	15 - 30 cm	132.06	135.83	19.24	496.00
PH	0 - 15 cm	7.00	0.36	6.47	8.08
	15 - 30 cm	7.00	0.37	6.38	7.97
Available P (ppm)	0 - 15 cm	31.94	33.54	7.92	137.19
	15 - 30 cm	37.17	66.87	3.75	271.88
Ca	0 - 15 cm	21.54	8.37	7.05	36.33
Meq/100gm	15 - 30 cm	21.87	8.31	9.48	37.29

the same with both soil depths while organic carbon and EC had an opposite trend (Table 2). This is consistent with finding for some Vertisols of Bale Mountain [1] .

The decline in soil carbon over the past two decades was attributed due to decrease in soil nutrient cycling under mixed crop-livestock production systems. This has happened for the reason that straw and crop stalks are used as animal feed, for construction material and off-farm sales. In addition, dung was used as fuel for household energy consumption and sold for income generation instead of being used as organic fertilizer [27] . In general there is a net export of nutrients from farm soil that is imbalanced with import.

In the present study about 94% of the arable land had a phosphorus deficiency (available P < 10 $mg \cdot kg^{-1}$), and 67% of that had a severe phosphorus deficiency (available P < 5 $mg \cdot kg^{-1}$) [29] . Therefore, large amounts of phosphorus fertilizer inputs were the primary means to gain higher yields in agriculture in Ethiopia [27] . To compensate for the deficiency high amounts of phosphorous were applied to soils to get better crop yield. However, excessive and inappropriate use of phosphorus fertilizer would inevitably lead to chemical and physical changes in soils [26] [30] . On top of this the characteristics of moderate CEC and low SOC content in the study area could made unfavorable for heavy metal sorption [12] . As a result the availability of heavy metals to plants nutrition would be in critical condition possibility due to Phosphorous over fertilization and lower soil organic carbon content [31] has reported that grain and straw yield of teff crop in central Ethiopia highland has quadrupled when supplemented by micro-nutrients such as Zn, Cu, Mg and S.

DTPA extractable critical concentrations for plant growth were 0.12 to 0.25 mg/kg for copper, 2.5 to 5.0 mg/kg for iron, 1.0 to 5.0 mg/kg for manganese and 0.5 to 1.0 mg/kg zinc [1] . DTPA extracted micro-nutrients of this study were 40.5 mg/kg for copper, 35.5 mg/kg for iron, 134 mg/kg for manganese, 2.1 mg/kg for nickel, 2.6 mg/kg for lead and 53.5 mg/kg for zinc. All the available micronutrients concentration for the central Ethiopian highland soils seems to be adequate for plant growth (Table 3). In terms of toxicity, the values of Cu, Fe, Mn, Zn, Ni and Pb found in the present investigation were lower than those of above critical value set by the international guidelines and other researchers (Table 3).

3.2. Relationship between Heavy Metals and Different Organic Sources

In physical terms organic matter improves the stability of soil aggregates resulting in better aeration of the soil and better water retention. It also improves the hydraulic conductivity as a result of the increasing percentage of soil macro-pores [32] . Thus, it contributes to the increase in adsorption.

Fulvic acids are organic substances found in organic compounds in soils, which predominantly have lower molecular weights than humic acids, usually not more than 5,000 gm mol^{-1}, and are often called dissolved organic carbon (DOC) [33] . These low molecular weight materials easily dissolve in aqueous media in the range of pH normally found in soils and that they can be taken up by plant roots directly along with the metals that they have bound [34] [35] .

In the current study the relationship between organic sources and trace minerals was identified as substantial. It was found that Cu, Zn and Ni have significant differences ($P < 0.05$) between different organic sources at 0 to 15 cm soil depth. In case of Zn there was a significant difference observed between animal sources (0.86 ppm ± 0.58) and the other two organic sources (no fertilizer addition 0.44 ppm ± 0.15 and plant sources 0.42 ppm ± 0.17). It showed that animal source was the major contributor to Zn load at 0 to 15 cm soil depth. Ni was higher for non-organic sources as shown in Table 4. For Cu there was no significant difference between animal sources (0.50 ppm ± 0.43) and no fertilizer added (0.51 ppm ± 0.22). There was a significant difference in Cu content between plant sources (0.18 ppm ± 0.09); and animal sources and no fertilizer addition at 0 to 15 cm depth, which indicated that the contribution of plant debris to Cu stock in the soil was low. There was no significant difference ($P < 0.05$) for Pb, Mn and Fe at 0 to 15 cm soil depth. In general at a soil depth of 0 - 15 cm the concentration of all heavy metals was higher for those organic sources from animal origin than for those from plant origin (see Figure 3 and Table 4).

Table 3. Comparison between heavy metal loads in the study area and other locations.

Element	Bulbula (Addis Ababa)	Kera (Addis Ababa)	Current study area	Guidelines for metal above standard level
	Soil (total) ($mg \cdot kg^{-1}$)	Soil (total) ($mg \cdot kg^{-1}$)	Soil (total) ($mg \cdot kg^{-1}$)	Soil (total) ($mg \cdot kg^{-1}$)
Cu	38.96	55	40.5	100
Fe	163.86	79.7	35.5	50000
Mn	6587	3598	134.0	2000
Ni	74.13	115	2.1	50
Pb	46.47	110	2.6	100
Zn	2985.5	263	53.5	300

Source: For Bulbula and Kera [2] [12] and for Guidelines [33]

The heavy metal concentration was significantly influenced by the addition of fertilizers from different organic sources, especially for Fe and Zn at 15 to 30 cm soil depth. The F test at $P < 0.05$ based on the linearity of independent variables for paired comparisons among the estimated marginal means of Fe and Zn showed a significant difference in the addition of fertilizer from organic sources into the soil. The mean Fe concentration at 15 to 30 cm depth was 8.88 ppm ± 8.95 for animal sources as compared from plant sources of 1.54 p.m. ± 2.32 and no-fertilizer was 2.41 ppm ± 1.76.

At the depth of 15 to 30 cm, the concentration of Zn was higher for Animal sources (0.86 ppm ± 0.58) medium for no fertilizer (0.44 ppm ± 0.15) and lower for plant sources (0.42 ppm ± 0.17). The result showed at 15 - 30 cm soil depth implied that animal waste products had a significant impact on the concentration of Fe and Zn in the subsoil. Other trace minerals Cu, Mn, Ni and Pb did not show any significant difference for different organic sources at 15 to 30 cm depth, particularly the insignificance of Pb could be associated with poor mobility potential of Pb in the soil [36] [37] .

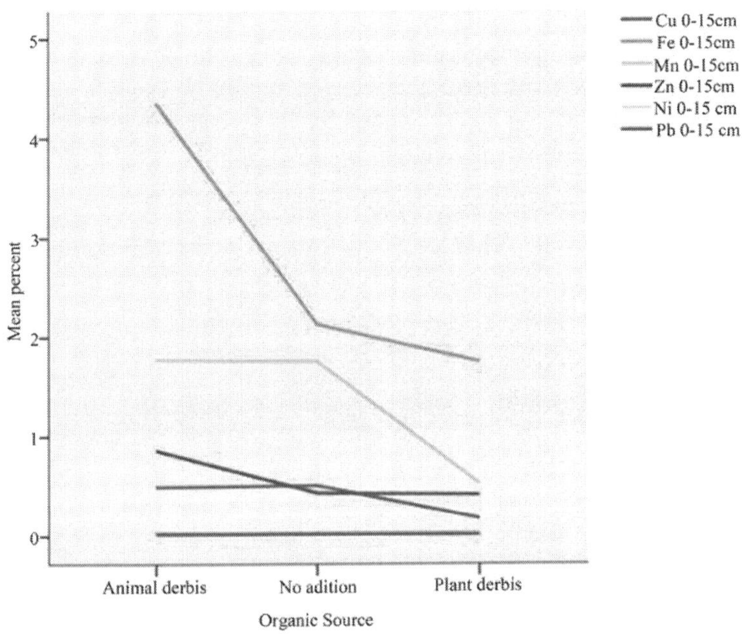

Figure 3. The mean percent of heavy metals by organic sources at 0 - 15 cm soil depth.

Table 4. Heavy metal concentration by organic sources.

		Organic source		
Element	Soil depth	Animal source (N = 5)	No organic fertilizer (N = 6)	Plant source (N = 4)
Cu	0 - 15 cm	0.50 ± 0.43	0.51 ± 0.22	0.18 ± 0.09
	15 - 30 cm	0.50 ± 0.51	0.46 ± 0.28	0.15 ± 0.09
Fe	0 - 15 cm	4.36 ± 4.32	2.13 ± 1.37	1.76 ± 2.16
	15 - 30 cm	8.88 ± 8.95	2.41 ± 1.76	1.54 ± 2.32
Mn	0 - 15 cm	1.78 ± 2.71	1.76 ± 0.43	0.53 ± 0.35
	15 - 30 cm	1.60 ± 2.01	1.46 ± 0.42	0.48 ± 0.36
Zn	0.15 cm	0.86 ± 0.58	0.44 ± 0.15	0.42 ± 0.17
	15 - 30 cm	0.75 ± 0.51	0.04 ± 0.01	0.38 ± 0.19
Ni	0 - 15 cm	0.01 ± 0.01	0.04 ±0.01	0.01 ± 0.01
	15 - 30 cm	0.02 ± 0.01	0.03 ± 0.01	0.01 ± 0.00
Pb	0.15 cm	0.03 ± 0.01	0.02 ± 0.01	0.01 ± 0.01
	15 - 30 cm	0.02 ± 0.01	0.03 ± 0.01	0.02 ± 0.01

In general animal source contribution to heavy metal concentration was higher for Zn and Fe at both 0 to 15 cm and 15 to 30 cm soil depth as shown in Table 4 and Figure 3, but still their level of concentration was below the environmental hazard standard this is mainly attributed due to high organic carbon in animal sources than others.

In the present study animal sources enhanced heavy metal mobility down the soil profile (except for Pb) when compared with two other sources, mainly due to relatively high organic carbon content from animal sources (1.96% organic carbon (OC) ± 1.32 at 0 to 15 cm depth and 1.57% OC ± 0.73 at 15 to 30 cm depth). Carbon content of plant sources was 1.20% OC ± 0.92 at 0 to 15cm depth and 0.88% OC ± 0.42 at 15 to 30 cm, and that of no-fertilizer was 1.06% OC ± 0.76 at 0 to 15 cm depth and 0.57% OC ± 0.64 at 15 to 30 cm. These findings were in agreement with the findings of [33] , who reported that the addition of high carbon sewerage sludge enhanced trace elements mobility down the soil profile in Greece and United Kingdom. Other study found that well decomposed, low molecular weight organic compounds may facilitate the transport of heavy metals into the soil profile [38] . Soil scientists tested this hypothesis and found that some DOC had an effect, but not with a very substantial contribution [39] . Although the present concentration level of heavy metal was not a serious problem in the study area, but certain soil parameters should be studied before organic compounds are applied into the soil to avoid any risks of future groundwater contamination. In general at a soil depth of 15 - 30 cm the concentration of all heavy metals was higher for those organic sources from animal origin than for those from plant origin (see Figure 4).

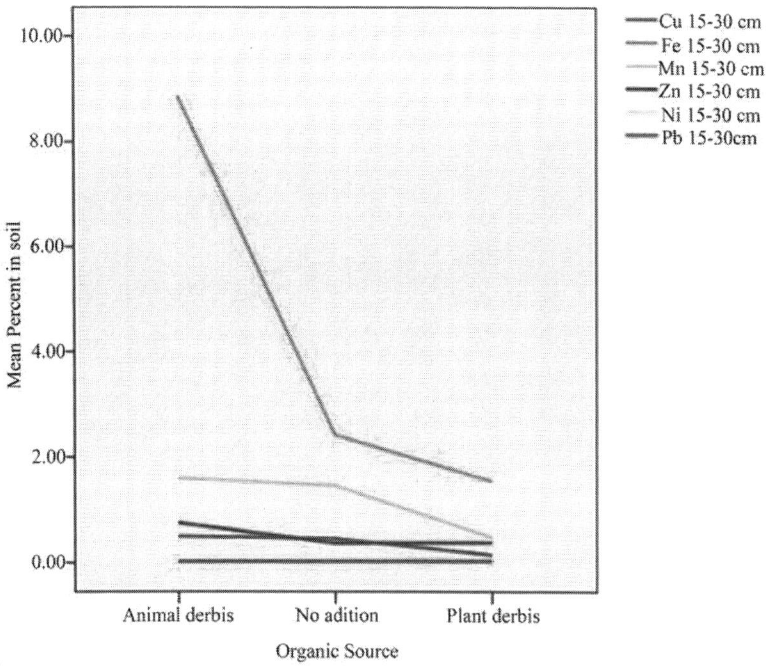

Figure 4. The mean percent of heavy metals by organic sources at 15 - 30 cm soil depth.

3.3. Heavy Metal Concentration and Land Use Type

Available heavy metals were the fractions of the total concentration of these metals in the soil, which was soluble in the soil solution or easily exchangeable from the soil surfaces. The availability of metals in soils depended on their forms and concentrations in the parent material and their input through commercial fertilizer, manure and municipal wastes, factory effluents and aerial deposition. In the absence of pollutants, natural levels of metals in soils were dictated by the types of elements in rocks, weathering rates, organic matter content, soil texture and soil depth [1] . It was also being affected by leaching processes, adsorption and desorption from the solid phase. Thus land use or management of a certain area has impacted on the heavy metal concentration of that specific locality [11] [40] .

In the current study relationship of heavy metal concentration and land use type gave a different picture for various land use categories as shown in Table 5 and Figure 4. There was a significant difference ($P < 0.05$) for Cu, Fe, Mn, Zn and Ni at 0 to 15 cm soil depth but no significant differences for Pb. Cu had a high significant difference between the sites for chemical discharges (0.83 ppm ± 0.27), cropland 0.28 ppm ± 0.14, grazing land 0.18 ppm ± 0.14 and Undisturbed site 0.34 ppm ± 0.12. This implied that Cu was one of the factory effluents discharged to farmers' fields that could pose a threat to future human and livestock health.

At the depth of 0 to 15 cm grazing land had a significant level of Fe (7.70 ppm ± 3.95) compared with other ecological groups (Cropland 0.82 ppm ± 0.81, chemical discharged sites 3.80 ppm ± 2.26 and undisturbed sites 2.60 ppm ± 1.86). The difference in Fe concentration could be due to livestock manure and urine on grazing land. Cropland had the lowest iron concentration of 0 to 15 cm depth, indicated that crops with high iron content like Teff [41] might have drained a significant amount of iron from the soil system, which mean that there was a need to compensate the soil for iron lost through nutrient absorption. High concentrations of Zn (1.26 ppm ± 0.90) and Mn (3.73 ppm ± 3.99) at 0 to 15 cm in undisturbed sites demonstrated that these two trace elements were depleted their concentration over time due to human anthropogenic factors.

There was a significant difference in Ni concentration (0.03 ppm ± 0.02) at 0 to 15 cm between chemical discharged sites and others land use groups which illustrated that effluents from factories with high Ni concentration could be a threat to farms situated near factories. In general chemical load by Ni and Cu at 0 to 15 cm depth was a threat, but still the hazard level is far below the environmental standard limit. In general farm sites near to chemical discharged areas have higher heavy chemical loads as compared to other site categories (see Figure 5 and Table 4).

There was a significant difference ($P < 0.05$) for Cu, Fe, Mn and Zn at 15 to 30 cm depth due to differences in ecological history. However, there was no significant difference for Ni and Pb at this soil depth. The concentration of Cu was significantly different between sites; chemical discharge (0.79 ppm ± 0.36), crop land (0.17 ppm ± 0.11) and grazing land (0.18 ppm ± 0.04), but there was no significant difference with undisturbed sites (0.58 ppm ± 042). At 15 to 30 cm depth the concentration of Fe was significantly different from crop land (0.77 p.m. ± 0.92) and the other three ecological groups (grazing land 6.30 ppm ± 1.85, undisturbed sites 10.30 p.m. ± 13.22 and sites with chemical discharges 6.63 ppm ± 6.76). It showed that crops absorbed a significant amount of Fe from the soil, particularly Teff for its highest iron content [41] .

Table 5. Heavy metal concentration in the soil by land use and soil depth.

		An ecological history			
Element	Soil depth	Chemical discharged sites (N = 4)	Crop land (N = 7)	Grazing land (N = 2)	Undisturbed sites (N = 2)
Cu	0 - 15 cm	0.83 ± 0.27	0.28 ± 0.14	0.18 ± 0.14	0.34 ± 0.12
	15 - 30 cm	0.79 ± 0.36	0.17 ± 0.11	0.18 ± 0.04	0.58 ± 0.42
Fe	0 - 15 cm	3.8 ± 2.26	0.82 ± 0.81	7.70 ± 3.95	2.60 ± 1.86
	15 - 30 cm	6.63 ± 6.76	0.77 ± 0.92	6.30 ± 1.85	10.3 ± 13.22
Mn	0 - 15 cm	1.70 ± 0.55	0.97 ± 0.72	0.26 ± 0.26	3.73 ± 3.99
	15 - 30 cm	1.54 ± 0.41	0.73 ± 0.52	0.08 ± 0.63	2.92 ± 3.01
Zn	0.15 cm	0.49 ± 0.16	0.43 ± 0.18	0.59 ± 0.15	1.26 ± 0.90
	15 - 30 cm	0.49 ± 0.20	0.35 ± 0.13	0.57 ± 0.13	0.98 ± 0.94
Ni	0 - 15 cm	0.03 ± 0.02	0.02 ± 0.01	0.01 ± 0.00	0.01 ± 0.00
	15 - 30 cm	0.03 ± 0.01	0.02 ± 0.01	0.00 ± 0.00	0.03 ± 0.02
Pb	0.15 cm	0.02 ± 0.01	0.02 ± 0.01	0.03 ± 0.01	0.03 ± 0.02
	15 - 30 cm	0.02 ± 0.01	0.02 ± 0.01	0.03 ± 0.01	0.03 ± 0.01

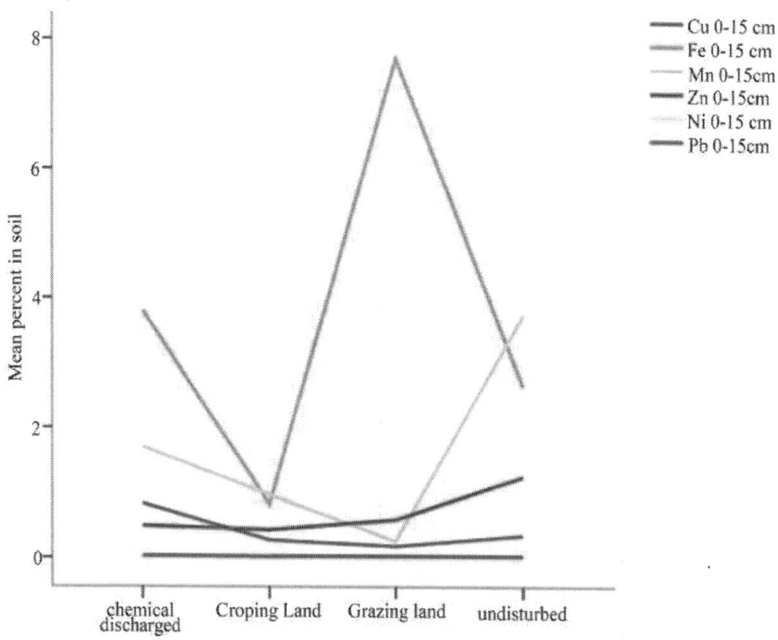

Figure 5. Micronutrients content with land use at 0 to 15 cm soil depth.

There was a significant difference ($P < 0.05$) for Cu, Fe, Mn and Zn at 15 to 30 cm depth due to differences in ecological history. However, there was no significant difference for Ni and Pb at this soil depth. The concentration of Cu was significantly different between sites; chemical discharge (0.79 ppm ± 0.36), crop land (0.17 ppm ± 0.11) and grazing land (0.18 ppm ± 0.04), but there was no significant difference with undisturbed sites (0.58 ppm ± 042). At 15 to 30 cm depth the concentration of Fe was significantly different from crop land (0.77 p.m. ± 0.92) and the other three ecological groups (grazing land 6.30 ppm ± 1.85, undisturbed sites 10.30 p.m. ± 13.22 and sites with chemical discharges 6.63 ppm ± 6.76). It showed that crops absorbed a significant amount of Fe from the soil, particularly Teff for its highest iron content [41] .

Mn concentration at the depth of 15 to 30 cm was significantly different between grazing land and the other three groups of ecological history. The highest Mn concentration was from undisturbed sites (2.92 ppm ± 3.01) and the lowest was from grazing land (0.08 ppm ± 0.63). Zn concentration was significantly different between the undisturbed sites and the other three ecologically different groups. There was a higher concentration of Zn at 15 to 30 cm depth for undisturbed sites (0.98 ppm ± 0.94) compared with the other three ecologically different groups as shown in Table 5 and Figure 6.

Crop lands suffer from low amounts of trace elements, which calls for further investigation to see how micro- nutrients could be supplemented to crop lands based on crop nutrient requirements. Some crops such as Teff remove more nutrients from the soil than others. This was in agreement with the findings by other workers who reported that Teff has high micro nutrient content demands such as Ca, Co, Zc, Al, Se and others [31] [41] .

This argument was further consolidated by findings obtained by [31] who reported that with space planting of 20 × 20 cm and with the addition of NPK plus Zn, Cu, Mn and other micro-nutrients, Teff yield increased to 8.7 tons/ha, compared with the control experiment which yielded 1.5 tons/ha. This was an increase by 6 folds, in contrast with that of average yield of 1.2 tons/ha by local farms. It was also reported that under the same experimental situation straw yield has increased to 25.8 tons/ha from the traditional 3 tons/ha after the applica- tion of macro and micro-nutrient elements compared with no fertilizer application, which yielded 9 tons/ha. It was concluded that Teff varieties responded well to micronutrients supplementation particularly those that contain micro-nutrients such as Zn, Cu, Mg and S. In general there was a higher concentration of heavy metal at 15 to 30 cm depth for farm lands with chemical discharged compared with the other three ecologically different groups as shown in Table 5 and Figure 6.

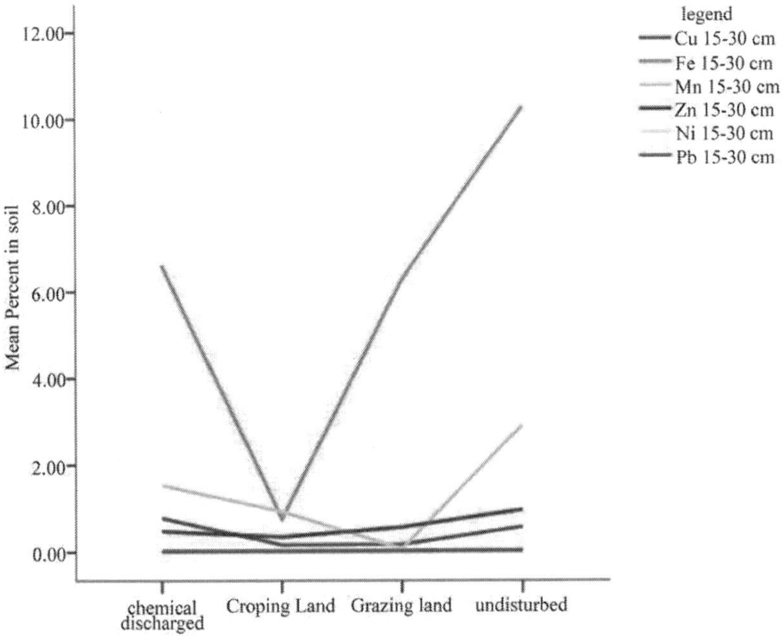

Figure 6. Micronutrients content with land use at 15 to 30 cm depth.

4. CONCLUSIONS

In terms of toxicity, the values of Cu, Fe, Mn, Zn, Ni and Pb found in the present investigation were lower than those of above critical value set by the international guidelines and other researchers. Although their impact was negligible, variability in heavy metal content was revealed mainly due to factory effluents discharge and organic sources disposal from animal products. The role of natural resources such as soil sediments, rocks and soil contents to heavy metals variability is limited.

Land use type contributed to the variability of heavy metals, particularly Fe which was very lower in cultivated soils than in an area that had never been cultivated. This showed that cereal crops like teff with higher Fe, Cu, Zn, Mn and other micro-nutrients content in their grains and straw are heavily mining these soil nutrients. However, all the available micronutrient concentrations for the central Ethiopian highland soils seem to be adequate for plant growth, but the availability of heavy metals to plants nutrition is in critical condition possibility due to higher phosphorous fertilization and lower soil organic carbon content. A report from soil scientists indicated that grain and straw yield of cereals like teff crop had over quadrupled when supplemented by micro-nutrients such as Zn, Cu, Mg and S.

In the present study, the content of organic carbon in the soil is declining rapidly that will create further obstruction to the availability of micronutrients to plants in the future. For better crop productivity, more organic matter should be added to the soils of central Ethiopian highland. The traditional nutrient cycling practices, such as manure application, allow crop residue to decay on fields, fallow and rotational cropping has to be followed to rehabilitate the soil and buildup both macro-and micro-nutrients to their natural level.

ACKNOWLEDGEMENTS

We thank Mengistu Asfaw for payment of publication processing fee and Yasin Getahun for mapping sample sites.We gratefully acknowledge International Livestock Research Institute(ILRI) for supporting this research.

REFERENCES

1. Yerima, B.P.K., Van Ranst, E., Sertsu, S. and Verdoodt, A. (2013) Pedogenic Impacts on the Distribution of Total and Available Fe, Mn, Cu, Zn, Cd, Pb and Co Contents of Vertisols and Vertic Inceptisols of the Bale Mountain Area of Ethiopia. African Journal of Agricultural Research (AJAR), 8, 5429-5439.
2. Itanna, F. (1998) Comparative Study on Soil Pollution with Toxic Substances on Farm Lands Close to Old and New Industrial Sites in Ethiopia. Bulletin of the Chemical Society of Ethiopia, 12, 105-112.
3. Alemayehu, T. (2001) The Impact of Uncontrolled Waste Disposal on Surface Water Quality in Addis Ababa. Ethiopian Journal of Science, 24, 93-104.
4. Itanna, F. (2002) Metals in Leafy Vegetables Grown in Addis Ababa and Toxicological Implications. Ethiopian Journal of Health Development, 16, 295-302.

5. Itanna, F. and Olsson, M. (2004) Land Degradation in Addis in Ababa Due to Industrial and Urban Development. Ethiopian Journal of Developmental Research, 26, 77-100.

6. Alemayehu, T. (2006) Heavy Metal Concentration in the Urban Environment of Addis Ababa, Ethiopia. Soil and Sediment Contamination, 15, 1-12.

7. Hailu, A. (2007) Metal Concentration in Vegetable Grown in Northern Addis Ababa, Ethiopia. MSc Thesis, School of Graduate Study, Addis Ababa University, 79.

8. Weldemariam, M.M. (2014) Heavy Metal (Pb, Cd, Zn, Cu, Cr and Fe) Content in Almeda Textile Industry Sludge, Northern Tigrai, Ethiopia. International Journal of Scientific and Research Publications (IJSRP), 4, No. 1.

9. Karathanasis, A.D. and Seta, A.K. (1993) Background Levels of Heavy Metals in Some Kentucky Soils. Agricultural Experimental Station, College of Agriculture, Bull 727 Univ. of Kentucky.

10. Esser, K.B., Bockheim, J.G. and Helme, P.A. (1991) Trace Element Distribution in Soils Formed in the Indiana Dunes. United States of America Soil Science, 152, 340-350.

11. Lee, B.D., Carter, B.J., Basta, N.T. and Weaver, B. (1997) Factors Influencing Heavy Metal Distribution in Six Oklahoma Benchmark Soils. Soil Science Society of America Journal, 61, 218-223.

12. Kabata-Pendias, A. (2007) Trace Elements in Soils and Plants. 3rd Edition, CRC Press, Boca Raton, FL, USA.

13. Kahsay, B. (2004) Land Use and Land Cover Changes in the Central Highlands of Ethiopia: The Case of Yere Mountains and Its Surroundings. M.Sc. Thesis, Addis Ababa University, 147.

14. Minase, N.A. (2014) Assessment of Environment-Livestock Interactions in Crop-Livestock Production System of the Central Ethiopian Highland. PhD Thesis, University of South Africa, Pretoria, RSA.

15. Tadesse, T., Haque, T. and Aduayi, E.A. (1991) Soil, Plant, Water, Fertilizer, Animal Manure and Compost Analysis Manual. ILCA PSD Working Document, B13, International Livestock Center for Africa, Addis Ababa, Ethiopia.

16. Olsen, S.R., Cole, C.V., Watanabe, F.S. and Dean, L.A. (1954) Estimation of Available Phosphorus in Soils by Extraction with Sodium Bicarbonate. Circular/United States Department of Agriculture, No. 939.

17. Walkley, A. and Black, I.A. (1934) An Examination of the Degtjareff Method for Determining Soil Organic Matter and a Proposed Modification of the Chromic Acid Titration Method. Soil Science, 37, 29-38.

18. Chapman, H.D. (1965) Cation Exchange Capacity. In: Black, C.A., Ed., Methods of Soil Analysis, American Society of Agronomy, Madison, 891-901.

19. SPSS (Statistical Package for Social Sciences) (2008) SPSS Version 17.0.1 for Windows. SPSS Inc., USA.

20. Everitt, B.S., Landau, S. and Leese, M. (2001) Cluster Analysis. 4th Edition, Arnold, London.

21. Manly, B.F.J. (2005) Multivariate Statistical Methods: A Primer. 3rd Edition, Chapman and Hall, London.

22. Zarcinas, B.A., Ishak, C.F., McLaughlin, M.J. and Cozens, G. (2004) Heavy Metals in Soils and Crops in Southeast Asia. 1. Peninsular Malaysia. Environmental Geochemistry and Health, 26, 343-357.

23. Alloway, B.J. (1995) Soil Processes and the Behavior of Metals. In: Alloway, B.J., Ed., Heavy Metals in Soils, Blackie Academic & Professional, London, 38-57.

24. Förstner, V. (1995) Riverine and Estuarine Sedimentation of Pollutants and Leaks from Sludges and Wastes: Analytical, Prognostic and Remedial Approaches. Land Degradation & Development, 4, 281-296.

25. Rodriguez, J.A., Nanos, N., Grau, J.M., Gil, L. and Lopez-Arias, M. (2008) Multi-Scale Analysis of Heavy Metal Contents in Spanish Agricultural Top Soils. Chemosphere, 70, 1085-1096.

26. Cheng, S. (2007) Heavy Metal Pollution in China: Origin, Pattern and Control: A State-of-the-Art Report with Special Reference to Literature Published in Chinese Journals. Environmental Science and Pollution Research, 14, 489-489.

27. Haque, I., Abebe, M., Mamo, T. and Dibabe, A. (1993) Nutrient Management. Improved Management of Vertisols for Sustainable Crop-Livestock Production in the Ethiopian Highlands: Synthesis Report 1986-92. Technical Committee of the Joint Vertisol Project, ILCA, Addis Ababa, 51-74.

28. Mamo, T., Haque, I. and Kamar, C.S. (1988) Phosphorus Status of Some Ethiopian Soils. In: Jutzi, S.C., Haque, I., McIntire, J. and Stares,

J.E.S., Eds., Management of Vertisols in Sub-Sharan Africa: Proceedings of a Conference Held at ILCA, Addis Ababa, Ethiopia, 31 August-4 September 1987, ILCA, Addis Ababa, 232-252.

29. Mamo, T. and Haque, I. (1987) Phosphorus Status of Some Ethiopian Soils. I. Sorption Characteristics. Plant and Soil, 102, 261-266.

30. Shepherd, J.G., Kleemann, R., Bahri-Esfahani, J., Hudek, L., Lalith, S.L., Elke, V.E. and van Dijk, K.C. (2016) The Future of Phosphorus in Our Hands. Nutrient Cycling in Agroecosystems, 104, 281-287

31. Berhe, T. (2010) Increasing Productivity of Teff: New Approaches with Dramatic Results. Report, Sasakawa Africa Association, Regional Office, Addis Ababa.

32. Chaney, R.L. and Ryan, J.A. (1993) Heavy Metals and Toxic Organic Pollutants in MSW-Compost: Research Results on Phytoavailability, Bioavailability, Fate, etc. In: Hoitink, H.A.J. and Keener, H.M., Eds., Science and Engineering of Composting Design, Environmental, Microbiological and Utilization Aspects, Renaissance Publications, Worthington, 451-506.

33. Antoniadis, V. (1998) Heavy Metal Availability and Mobility in Sewage Sludge Treated Soils. Ph.D. Thesis, Department of Soil Science, the University of Reading, Reading, 255 p.

34. Hamon, R.E., Lorenz, S.E., Holm, P.E., Christensen, T.H. and McGrath, S.P. (1995) Changes in Trace Metal Species and Other Components of the Rhizosphere during Growth of Radish. Plant, Cell & Environment, 18, 749-756.

35. Alloway, B.J. (1994) Toxic Metals in Soil-Plant Systems. John Wiley and Sons, Chichester.

36. Williams, D.E., Vlamis, J., Pukite, A.H. and Corey, J.E. (1980) Trace Element Accumulation, Movement and Distribution in the Soil Profile from Massive Applications of Sewage Sludge. Soil Science, 129, 119-132.

37. Mogo, F.C. (2002) Assessment of Heavy Metals in Soil, Water, Sediments and Certain Plant Taxa in Imo and New Calabar Rivers. Ph.D. Thesis, University of Port Harcourt, Port Harcourt.

38. Jardine, P.M., Dunnivant, F.M., Selim, H.M. and McCarthy, J.F. (1992) Comparison of Models for Describing the Transport of Dissolved Organic Carbon in Aquifer Columns. Soil Science Society of America Journal, 56, 393-401.

39. Guisquiani, P.L., Gigliotti, G. and Businelli, D. (1992) Mobility of Heavy Metals in Urban Waste-Amended Soils. Journal of Environmental Quality, 21, 330-335.

40. He, Z.L.L., Yang, X.E. and Stoffella, P.J. (2005) Trace Elements in Agroecosystems and Impacts on the Environment. Journal of Trace Elements in Medicine and Biology, 19, 125-140.

41. Wondimu, A. and Tekabe, F. (2001) Utilization of Teff in the Ethiopian Diet. Narrowing the Rift. Teff Research and Development: Proceedings of the International Workshop on Teff Genetics and Improvement, Debre Zeit, 16-19 October 2000, 239-244.

3

CHAPTER

Evaluation of Soil Water Availability (SWA) Based on Hydrological Modelling in Arid and Semi-Arid Areas: A Case Study in Handan City, China

Xuerui Gao [1,2,*], Jianhua Wang [3], Pute Wu [1,2], Yong Zhao [3], Xining Zhao [1,2] and Fan He [3]

[1]*Institute of Soil and Water Conservation, Northwest A&F University, Yangling 712100, Shaanxi, China*

[2]*Institute of Water Saving Agriculture in Arid Regions of China, Northwest A&F University, Yangling 712100, Shaanxi, China*

[3]*State Key Laboratory of Simulation and Regulation of Water Cycle in River Basin, China Institute of Water Resources and Hydropower Research, Beijing 100038, China*

ABSTRACT

To evaluate the spatial and temporal soil water availability (SWA) and improve the soil water use efficiency in agriculture, this study established a new index system for SWA evaluation considering the soil storage capacity, the temporal and spatial matching degree between soil water supply and crop water demand, and the soil water transformation and utilization efficiency. To quantitatively calculate the indexes, a self-developed hydrological model (MODCYCLE) was introduced and applied in Handan City. The results show that MODCYCLE performed well in simulating the soil hydrological processes. A comparison of the observed soil moisture data and the simulated values revealed relative errors below 10% and correlation coefficients all above 0.7. According to the evaluation results, the SWA for different typical years was very unevenly distributed in Handan City. The average SWA in a normal year is 0.71, which is higher than that in dry years at 0.62 and in wet years at 0.65. Regarding different regions, the average SWA was generally much higher in the eastern plain area than that in the western mountainous area. Generally, there is a great potential for improving SWA and further the sustainable development of soil water resource, particularly for dryland farming in similar areas.

Keywords: soil water resource; soil water availability (SWA); MODCYCLE; dryland farming; Handan City

1. INTRODUCTION

Influenced by both climate change and high-intensity human activities, the volume of runoff water resources is currently decreasing sharply [1]. Under this background, water use competition among different sectors (e.g., urban, industry, agriculture, and ecology) is increasing. The agriculture sector is known to be vulnerable compared with other sectors that have better economic benefits [2]. Therefore, agricultural water resources are largely diverted to other sectors to pursue higher benefits. This practice is definitely a great threat to grain production and food security. To guarantee sustainable agricultural development and to promote food security, an increasing number of studies have focused on soil water resources, particularly in arid and semi-arid areas. In general, water can be directly absorbed by crops and plants after it is transformed into soil moisture. Therefore, soil water, as a type of water resource, plays an important role in agricultural production and food security. However, agricultural specialists and water resource managers have traditionally paid much attention to the "visible" water, e.g., surface water and ground water, which is also known as blue water. Owing to the difficulty of monitoring and regulating soil water, the utilization value of soil water resources is largely ignored, particularly in Chinese agriculture. Recently, soil water monitoring and simulation technologies have developed significantly, which has provided a good foundation for the efficient use and management of soil water resources in rain-fed agriculture, particularly in arid and semiarid areas.

Soil moisture, known as a direct index of soil water availability (SWA), is an important indicator for farmland hydrological forecasts and irrigation scheduling. Therefore, most previous studies have focused on the monitoring and forecasting of soil moisture by using different techniques. Robinson et al. [3] concluded that there are two main approaches for studying the spatial–temporal distribution of soil moisture. The first method refers mainly to measurement methods by direct contact with the soil profile, such as gravimetric methods [4], tensiometers [5], the neutron probe method [6], and time domain reflectometry (TDR) [7,8]. Tanaka and Aase [9] used the neutron scattering technique to measure soil moisture at the field scale to determine the season in which the fallow method might influence soil water storage. Zegelin et al. [10] improved the field probes to detect the soil water content based on TDR, which has been widely used owing to its convenience and economic efficiency. Gao et al. [11] studied the spatial variability of the root zone soil moisture (RZSM) in gullies by using the observed soil moisture data collected by the TDR method from nearby uplands. The soil moisture data collected by the traditional direct methods are accurate but are time-consuming and cost intensive if the sampling area is large [12].

The second method, which refers mainly to remote sensing techniques [13], cannot directly measure the RZSM. However, it can sense the RZSM without direct contact with the soil based on multi-spectral information from the soil layer. Remote sensing methods include passive microwave radiometers, synthetic aperture radars, scatterometers, and thermal methods. These methods are either ground based or operated from airborne or spaceborne platforms [14]. Recent progress in remote sensing technologies has allowed us to obtain frequent soil moisture distribution maps easily and quickly [15,16]. Kong and Dorling [17] retrieved near-surface soil moisture from advanced synthetic aperture radar imagery, which showed satisfactory accuracy for the shallow soil layer profile. Ford et al. [18] used an exponential decay filter to estimate RZSM using the near-surface soil moisture derived from the Soil Moisture and Ocean Salinity (SMOS) satellite. Martínez-Fernández et al. [19] calculated the soil water deficit index (SWDI) based on SMOS satellite remote sensing data for agricultural drought monitoring. The advantage of remote sensing is that it gives uniform spatial and temporal coverage of the estimated soil moisture easily and quickly. However, the limitations include confinement to the top few centimetres of the soil layer and coarse spatial and temporal resolutions. However, when integrated with a land surface process model, remote sensing can estimate the soil moisture both at the surface and deep in the root zone with high spatial and temporal resolutions. In addition, data assimilation provides a means for combining the advantages of modelling with those of remote sensing data. A host of studies based on remote sensing measurement have used data assimilation to derive an indirect estimate of the RZSM [20]. Montaldo and Albertson [21] developed a multi-scale assimilation approach for RZSM estimation, which is shown to provide marked improvements. Reichle and Koster [22] assimilated the Scanning Multi-channel Microwave Radiometer (SMMR)-derived real surface soil moisture data into the National Aeronautics and Space Administration (NASA) Catchment land surface model at a large continental scale over a long period. The soil moisture estimates in this research are significantly better than those obtained with the model or from the satellite alone. Moreover, using statistical and distributed hydrological models to simulate the spatial and temporal pattern of soil water has become an attractive method because it is more time saving and cost effective than traditional methods. Kim [23] developed a time series model for soil moisture simulation and the results showed that the simulated values fit well with the observed soil moisture series. Gao et al. [24] mapped the spatial and temporal distribution of soil water moisture based on a self-developed hydrological model in Northern China. The physical-based distributed hydrological model showed good reliability for soil moisture simulation.

The studies mentioned above focus mainly on soil water observation, soil moisture mapping, and agricultural drought monitoring. Few scientists have sought to improve soil water use efficiency based on soil water content observation and simulation for agricultural water management. In the context of climate change and strong anthropogenic influence on the hydrological cycle, the available volume of blue water resources is shrinking significantly. Therefore, improving the utilization efficiency of green water resources and making full use of soil water resources in agriculture are of great concern, particularly in arid and semi-arid areas. It is widely accepted that evaluation of soil water resources availability is the first and most important step for high-efficiency soil water resources development and sustainable utilization. Therefore, the objective of this study is to introduce a distributed hydrological model to depict the soil water transformation and flow routes at a regional farmland scale. The soil water availability assessment index system is established to jointly consider factors such as soil storage capacity, soil water recharge and discharge rates, and soil water uptake efficiency by crops and plants. The main objectives of this study include (1) development of a spatially distributed hydrological model to simulate the soil water transformation and flow process based on the complex tillage and water management regimes of farmland in the North China Plain; (2) establishment of the soil water availability assessment index framework considering such parameters as soil properties, soil water reservoir characteristics, and soil water recharge and discharge efficiency; and (3) selection of Handan City in the North China Plain in a semi-arid area as the study region and quantitative calculation of the farmland soil water availability. In addition, we reveal the spatial and temporal variability of soil water use potential and efficiency in accordance with the evaluation results, and we provide specific and differential suggestions and strategies for different sub-regions in the study area. This study can serve as a guide for improving soil water use efficiency and maintaining sustainable agricultural and ecological development in arid and semi-arid areas.

2. MATERIALS AND METHODS

2.1. Soil Water Simulation Tool

Before SWA evaluation is conducted, the scientist should understand the soil water movement process in the farmland that is used to quantify the water availability index. However, for a large-scale basin or region, it is impossible to explore the soil water movement process though field experiments, which are time consuming and cost intensive. With the development of computer numerical simulation technology, the soil transformation and flow process at a large scale has been efficiently simulated by hydrological models in many studies [25]. In this study, the object-oriented modularised model for basin-scale water cycle simulation (MODCYCLE) is introduced to simulate the farmland hydrological processes and to map the temporal and spatial distribution of the farmland soil water moisture [24,26]. MODCYCLE, self-developed by the China Institute of Water Resources and Hydropower Research (IWHR), is a spatially distributed model properly combining a hydro-physical model with a conceptual model. Moreover, MODCYCLE avoids the difficulty in parameterization and overload of computation.

MODCYCLE is similar to the commonly used soil and water assessment tool (SWAT) model [27], particularly in its hydrological modelling and calculation. Both are semi-distributed and conceptual-physical models in which the main hydrological processes such as overland runoff, groundwater runoff, evapotranspiration, soil water infiltration, and percolation are all depicted with generalised physical methods and simplified empirical formulas. Numerous previous studies have proved that SWAT is an effective tool for simulating soil moisture distribution at the regional scale, particularly in natural basins or areas of limited anthropogenic influence [28,29]. However, it performs poorly when simulating the hydrological processes in areas influenced by strong human activities, especially for the complex cropping systems and irrigation regimes in China. Therefore, MODCYCLE is designed to improve the simulation accuracy under the conditions of strong anthropogenic influences and complicated Chinese tillage systems. According to the simulation principle, MODCYCLE uses the core of SWAT because the simulation principles of SWAT are proven to be effective and practical by a series of case studies and applications. Moreover, a special module, the agricultural management module, has been embedded in MODCYCLE to depict the complex hydrological processes of Chinese farmland with complicated tilling practices and strong anthropogenic influences. Based on the agricultural management module, an agricultural activity command database in accordance with the current farmland management regimes of China has been created, which controls MODCYCLE to precisely and elaborately simulate the farmland hydrological processes. In summary, compared with SWAT, MODCYCLE performs much better in depicting the farmland soil water movement processes in China. Therefore, it was used as an important tool for SWA evaluation in this study.

The structural framework of MODCYCLE is similar to that of SWAT, the most widely used hydrological model. The first step in MODCYCLE is to divide the study area into sub-basins. The sub-basins are further subdivided into different hydrological response units (HRUs). The runoff generated in each sub-basin is from two sources: the area within the sub-basin and the upstream contributing areas. Based on a digital elevation model (DEM) and GIS analysis, the topological relationships among the sub-basins are established, and the runoff flow directions in the sub-basin can be depicted. For SWA evaluation, soil water movement simulation is of extreme importance. Water molecules are absorbed in the soil mass forced by the soil matric potential in an unsaturated zone. Therefore, soil water moves much slower than surface water. However, compared with soil water movement in a horizontal direction, the speed in the vertical direction is much faster because of gravity. As a result, soil water movement in the horizontal direction in an unsaturated zone can be neglected, although that in the vertical direction must be considered. In MODCYCLE, the main hydrological process in the vertical direction is simulated through a one-dimensional (1D) Soil Column Method at the HRU scale (Figure 1).

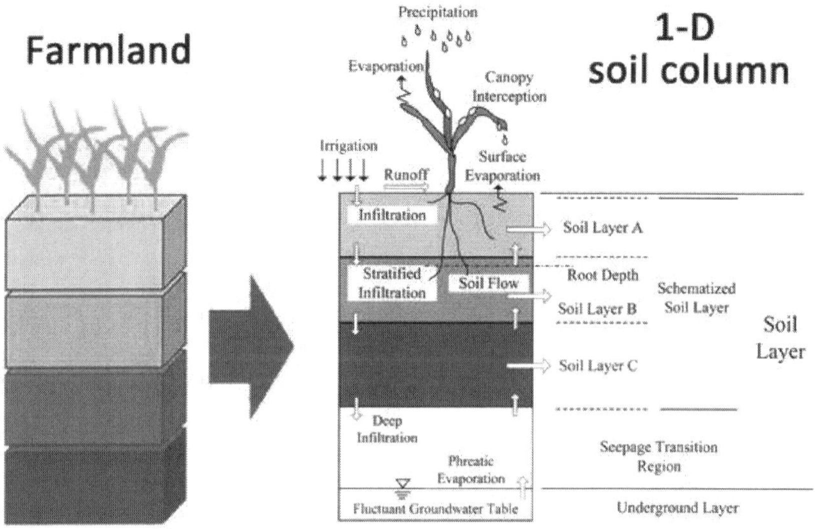

Figure 1. Hydrological processes in the 1D soil column method.

As shown in Figure 1, HRU is regarded as the basic simulation unit in the model. Vertical water movement consists of three processes: the surface water process, soil water process, and ground water process. Because the soil water process is the main concern in our study, the simulation method for soil water movement is introduced in detail below. In the 1D soil column, the soil layer is divided into two parts. The upper part is the schematised layer, and the lower part is the seepage transition layer, as shown in Figure 1. Ordinarily, the schematised layer is the crop root zone in farmland, where the soil water content fluctuates frequently at different timescales, particularly in the growing season. To precisely simulate the soil water content variation process in the crop root zone, the schematised layer should be also subdivided into different sub-layers (e.g., Layer A, Layer B, Layer C, etc.). In MODCYCLE, the schematised layer can be subdivided into 10 multiple soil layers. The seepage transition layer is directly beneath the schematised layer, where the soil water content is relatively stable. However, the seepage transition layer is an important linkage zone for water exchange between the ground water and soil water. That is, the seepage transition layer is a buffer zone in which the surplus water can be stored during the wet season and is supplied into the soil root zone for crop growth during the dry season. Therefore, the seepage transition layer plays a significant role in farmland cultivation and management. In the 1D soil column method, the water balance equation can be described as

$$\Delta SW = R + IR + GEV - SF - \Delta LF - UEV - PEV - SEV - SEP \quad (1)$$

where ΔSW is the change in the soil water content in the simulation period, R is the equivalent precipitation depth adjusted for canopy interception, IR is the irrigation depth in the simulation period, GEV is the evaporation depth from phreatic water to the crop root zone, SF is the surface runoff loss depth, ΔLF is the change in the surface water ponding depth, UEV is the evaporation depth from the surface water, PEV is the crop evapotranspiration depth, SEV is the evaporation depth from the soil layer, and SEP is the water deep infiltration depth from the soil layer to the groundwater aquifer; all are in millimetres. As described in Equation (1), the entire soil layer can be considered as a water tank. Most processes of water exchange in the soil layer occur through the top boundary (farmland surface) and the bottom boundary (ground water table). Nine variables need to be determined to solve the equation. Rand IR are input parameters in the model, and the calculation methods for the rest of the parameters can be obtained from related studies [26,30].

2.2. Soil Water Availability Evaluation Method

2.2.1. Definition of Soil Water Availability

Improving the efficiency of agricultural water resources is the central topic for water-saving agriculture technologies. In arid and semi-arid areas, any type of water resource should be transformed into soil water before it is effectively absorbed and utilised by crops and plants. Based on the analysis of soil water resource properties, this study argues that three factors will primarily influence soil water use efficiency at the farmland scale. The first is availability of the soil water quantity, which is mainly affected by soil particles and root zone depth. Not all water stored in the soil layer can be directly absorbed by crops; essentially, only the soil moisture above the wilting point and below the field capacity can be used by crops. Here, we introduce two first-class indices to represent the soil water quantity: soil water storage capacity [31] and the operational characteristics of the farmland soil reservoir [32], the calculation methods of which will be introduced later. The second factor is the matching degree of soil water resources between the supply and demand. In some cases, although there is sufficient soil water supply throughout the year, the crops may also suffer from drought in some months (e.g., April and May) because the temporal distribution of available soil water is uneven. Here, we introduce one first-class index to represent the matching degree of the soil water resource, which includes two sub-indices: the temporal matching degree index and the spatial matching degree index. The third factor is the efficiency of soil water transformation and uptake by plants. Here, we introduce one first-class index to represent the soil water efficiency [33,34], which also includes two sub-indices: soil water conversion efficiency and crop absorption efficiency. The details of the aforementioned indices are given in Table 1.

Table 1. Evaluation index system for farmland soil water availability.

Factor	First-Class Index	Second-Class Index	Description
Availability of soil water quantity	The soil water storage capacity (C1)	The valid capacity of soil reservoir (C11)	The valid water supply capacity (equivalent water depth) of the soil in the crop root zone, mm.
	Operational characteristics of farmland soil reservoir (C2)	Average value of annual empty storage capacity of the soil reservoir (C21)	To characterise the average soil water shortage severity in the soil root zone within a year, mm.
		Average annual variation of empty storage capacity of the soil reservoir (C22)	To characterise the fluctuation of water shortage severity in the root zone within a year, dimensionless.
The spatial and temporal matching of soil water resource between supply and demand	Temporal and spatial matching degree (C3)	Temporal matching degree index (C31)	The temporal matching degree between soil water supply and crop water requirement, dimensionless.
		Spatial matching degree index (C32)	The spatial matching degree between soil water supply and crop water requirement, dimensionless.
The efficiency of soil water transformation and plant uptake	Soil water transformation and crop absorption efficiency (C4)	Soil water conversion efficiency (C41)	To characterise the efficiency of water transforming from other types to soil water, dimensionless.
		Crop absorption efficiency (C42)	To characterise the efficiency the soil water is absorbed by crops and plants, dimensionless.

Previous studies have largely examined how to increase the efficiency of soil water transformation and plant absorption. However, they neglect to discuss the importance of the availability and quantity of soil water to be used by plants and crops and the degree of temporal and spatial matching degree between the soil water supply and the demand. In this study, we comprehensively consider these three factors and suggest a new concept for SWA evaluation. An optimal objective function is presented to quantitatively characterise the soil water availability at the farmland scale, as shown in Equation (2):

$$U = F(A, R, E) \tag{2}$$

where U represents the comprehensive SWA, F(X,Y,Z) is the soil water availability evaluation function, A represents the factor for the degree of difficulty of the soil water to be used by plants and crops, R represents the factor for the degree of matching between the soil water supply and the water demand, and E represents the factor for the efficiency of soil water transformation and plant uptake.

2.2.2. Framework of the Soil Water Availability Index System

As shown in Equation (2), the optimal objective function of soil water availability is only a conceptual expression and it cannot be used to quantitatively evaluate the soil water availability directly. Therefore, an index system was introduced to quantitatively characterise the three factors. Based on the index system, we used the analytic hierarchy process (AHP) method to calculate the comprehensive SWA.

The SWA index system, presented in Table 1, includes four first-class indices and seven second-class indicators. The first and second first-class indices (C1 and C2) mainly represent the factor for the degree of difficulty for the soil water to be used by plants and crops, which are further subdivided into three second-class indicators. The third first-class index (C3) represents the matching degree between the soil water supply and the water demand, which is further subdivided into two second-class indicators. The fourth first-class index (C4) represents the factor for the efficiency of the soil water transformation and plant absorption, which is further subdivided into two second-class indicators.

Index for Soil Water Storage Capacity (C1)

Soil water storage is a key parameter for farmland hydrology in arid and semiarid areas. In view of the water holding ability of the soil, scientists always regard this capacity as the soil water reservoir capacity. According to previous studies [31], soil water reservoir capacities include mainly the total soil water reservoir capacity (TC), soil water storage capacity (SC), unavailable soil water reservoir capacity (UC), available soil water reservoir capacity (AC), and soil dead water storage capacity (DC). In this study, we present a second-class index (C11) to represent the first-class index soil water storage capacity (C1). Here, C11 represents the available soil water reservoir capacity, which is essentially determined by the soil properties in the root zone of the farmland. With a total of n soil layers in the root zone, the valid capacity of the soil reservoir can be represented as

$$W_p = \sum_{i=1}^{n} \left(FC\,(i) - \theta_{wp}\,(i)\right) D_i \tag{3}$$

where Wp represents the valid storage capacity (equivalent water depth) of the soil water in the crop root zone; FC(i) represents the field soil water holding capacity (equivalent water depth) of the soil layer i; θwp(i) is the volume of the soil water (equivalent water depth) at the wilting point of the soil layer i, and Di represents the thickness of the soil layer i, all of which are in millimetres. In addition, nrepresents the number of soil layers in the root zone.

Index for the Operational Characteristics of the Farmland Soil Reservoir (C2)

To characterise the average soil water shortage severity in the soil root zone within one year, the two second-class indices (C21 and C22) are introduced to represent the index for the operational characteristics of the farmland soil reservoir. The index C21 is the average value of the annual empty storage capacity of the soil reservoir:

$$\overline{W_e} = \left(\sum_{j=1}^{m} \sum_{i=1}^{n} \left(\left(FC\,(i) - \theta_j\,(i) \right) D_i \right) \right) / m \tag{4}$$

where $\overline{W_e}$ is the average value of the annual empty storage capacity (equivalent water depth) of the soil reservoir, in millimetres; FC(i) represents the field soil water holding capacity (equivalent water depth) of the soil layer i, in millimetres; θwp(i) represents the actual soil water content (equivalent water depth) of the soil layer i; Di represents the thickness of the soil layer i, in millimetres; m represents the total number of days in the evaluation period; and n represents the number of layers of soil in the root zone.

The index C22 represents the average annual variation of the empty storage capacity of the soil reservoir for the root zone:

$$C_w = \sqrt{\frac{\sum_{j=1}^{m} \left(\sum_{i=1}^{n} \left(\left(FC\,(i) - \theta_j\,(i) \right) D_i \right) - \overline{W_e} \right)^2}{m}} / \overline{W_e} \tag{5}$$

where Cw is the average annual variation of the empty storage capacity of the soil reservoir, in millimetres. The other terms are the same as those described in Equation (4).

Index of Degree of Matching between the Soil Water Supply and Crop Water Demand (C3)

In this study, the two second-class indexes (C31 and C32) were originally introduced to characterise the matching degree between the soil water supply and crop water demand. The index C31 represents the temporal matching degree between the soil water supply and crop water demand:

$$m_{ti} = \begin{cases} e^{\frac{RC_i - DE_i - IR_i}{DE_i}}, & \text{if } RC_i - DE_i - IR_i < 0 \\ 1, & \text{if } RC_i - DE_i - IR_i \geq 0 \end{cases} \tag{6}$$

$$m_t = \frac{\sum_{i=1}^{n} m_{ti}}{n} \tag{7}$$

where mti represents the temporal matching degree between the soil water supply and crop water demand in month i in the growing period, with values ranging from 0 to 1; mt is the average temporal matching degree between the soil water supply and crop water demand in the entire growing period; and n is the number of months in the entire growing period for a specific planting structure. DEirepresents the total crop water demand in month i, and IRi represents the total artificial irrigation amount in month i, both in millimetres, because the concept of temporal matching degree between soil water supply and crop water demand does not consider the influence of artificial irrigation. Therefore, the irrigation amount should be excluded to calculate the index (C31). RCi is the total available soil water supply in month i, in millimetres. When RCi=ETi+RESi, where ETi is the total water evaporated from the crop and soil to the air in month i, RESi is the effective soil water remaining in the soil layer profile at the end of month i.

The index C32 represents the spatial matching degree between the soil water supply and crop water demand:

$$m_{sj} = \begin{cases} e^{\frac{RC_j - DE_j - IR_j}{DE_j}}, & \text{if } RC_j - DE_j - IR_j < 0 \\ 1, & \text{if } RC_j - DE_j - IR_j \geq 0 \end{cases} \tag{8}$$

$$m_s = \frac{\sum_{j=1}^{n} m_{sj} A_j}{A} \tag{9}$$

where msj represents the spatial matching degree between the soil water supply and crop water demand in the sub-region j in the growing period, with values ranging from 0 to 1; Aj is the area of the sub-region j, in square kilometres; A is the total area of the entire region, in square kilometres; and n is the number of sub-regions in the entire study region. DEj represents the total crop water demand in the sub-region jin the growing period, in millimetres; and IRj represents the total artificial irrigation amount for crop growth in the whole growing period in sub-region j. As in the temporal matching degree, the artificial irrigation amount should be excluded to calculate the index C32. RCj is the total available soil water supply in sub-region j in the entire growing period, in millimetres, which is described as RCj=ETj+RESj, where ETj is the total water evaporated from the crop and soil to the air in sub-region j in the entire growing period, in millimetres; and RESj is the effective soil water remaining in the soil layer profile in sub-region j at the end of the growing period, in millimetres.

Index for Soil Water Transformation and Crop Absorption Efficiency (C4)

The soil water transformation efficiency refers to the ratio between the effective soil water recharged in the root zone and the total volume of the other types of water (e.g., precipitation, irrigation water, and phreatic water) that can be transformed into soil water:

$$E_{ct} = \frac{\sum_{i=1}^{n}(\theta_i(t+1) - \theta_i(t))TH_i + ET(t) + ES(t) + R(t)}{NP(t) + I(t) + \varepsilon(t)} \quad (10)$$

$$E_c = \frac{\sum_{t=1}^{T} E_{ct}}{T} \quad (11)$$

where Ect is the daily soil water transformation efficiency during day t, which ranges from 0 to 1; Ecis the average soil water transformation efficiency in the entire growing period, which also ranges from 0 to 1; T is the total number of days in the entire growing period; θi(t+1) is the actual soil water content in soil layer i during day (t+1); θi(t) is the actual soil water content in soil layer i during day t; THi is the thickness of the soil layer i; ET(t) is the crop transpiration during day t; ES(t) is the soil water evaporation during day t; R(t) is the water loss through deep percolation during day t; NP(t) is the net precipitation adjusted for canopy interception; I(t) is the irrigation amount during day t and ε(t)is the volume of the ground water returned to the root zone through phreatic water evaporation during day t.

Based on the concept of water use efficiency (WUE) [34], this study proposes the index of the crop absorption efficiency to characterise the soil water use efficiency by crops and plants:

$$E_{dt} = \frac{ET(t)}{ET(t) + ES(t) + R(t)} \quad (12)$$

$$E_d = \frac{\sum_{t=1}^{n} E_{dt}}{n} \quad (13)$$

where Edt represents the soil water absorption efficiency by crops and plants in day t and Ed is the average soil water absorption efficiency in the entire growth period. The other terms are the same as those described previously.

2.2.3. Analytic Hierarchy Process

The AHP method was first developed in the 1970s by Saaty from the University of Pittsburgh [35]. Owing to its simplicity and great flexibility, the AHP method has been widely applied in engineering, management, manufacturing, marketing, and other applications [36]. Moreover, the AHP method has proven to be an efficient tool to help decision-makers achieve structured determinations, and it is used in multi-criteria evaluation for decomposing a complex and unstructured problem into its component parts and synthesizing the subjective judgment from experts into a hierarchical structure to determine the weights of different factors [37]. In this method, the structure usually has three or four levels. The highest level (I) reflects the overall goal of the problem. The second and third levels (II and III) include the factors and sub-factors, also known as criteria and sub-criteria, and the fourth level (IV) contains the decision options [38]. In this study, only three levels are used. The highest level is the overall goal, which is the total comprehensive soil water availability (SWA) for different regions. The second level includes the first-class indices (C1, C2, C3, and C4) from the index system, and the third level is composed of the second-class indices (C11, C21, C22, C31, C32, C41, and C42).

After the factor levels are established, the multi-criteria SWA evaluation using the AHP method can be conducted following four main steps.

First, a pair-wise comparison matrix is developed. The relative weight of importance or contribution of the factors (or sub-factors) with respect to an element in the upper level is evaluated via pair-wise comparison based on Saaty's nine-point scale [39]. Saaty's nine-point scale transforms the subjective judgment of the factors' importance into numerical values. Therefore, the matrix is generated by placing the factors (or sub-factors) horizontally and vertically. The values are defined by comparing the horizontal factor with the vertical factor. On the basis of this method, the pair-wise comparison matrix of the four factors in this study (C1, C2, C3, and C4) was determined, as shown in Table 2.

Table 2. Pair-wise comparison matrix of the factors (C1, C2, C3, and C4).

Factors	C_1	C_2	C_3	C_4
C_1	1	0.25	0.5	0.125
C_2	4	1	0.5	0.333
C_3	2	2	1	0.5
C_4	8	3	2	1

Secondly, to calculate the weights and contribution of the factors to the overall goal, we normalised the matrix and calculated the priority vectors. In our study, MATLAB software was used to calculate the priority vectors. According to the calculation, the priority vectors of the pair-wise comparison matrix separately were 4.17, 0.042, −0.106 + 0.842i, and −0.106 + 0.842i.

Thirdly, the consistency was checked. The reliability of the results needs to be tested by using the consistency ratio (CR), where CR=CI/RI. CI=(λmax−n)/(n−1), where CI is the consistency index; λmax is the maximum eigenvalue of the matrix, and λmax= 4.17 in this study; n is the number of factors; and RI is the random consistency index, which is the index of a pair-wise comparison matrix generated randomly according to the number of factors [40]. Normally, if CR is higher than 0.1, the consistency check is not acceptable. In this study, CR= (4.17 − 4)/(4 − 1)/0.9 = 0.063, which is <0.1. Therefore, the consistency is acceptable.

Finally, the normalised priority vector corresponding to the maximum eigenvalue of the pair-wise comparison matrix is determined as the weights of the different factors relative to the evaluation goal. According to the calculation through MATLAB, the priority vector W=(0.072, 0.183, 0.240, 0.505).
Thus, U=0.072·C1+0.183·C2+0.240·C3+0.505·C4, where, U is the comprehensive soil water availability; and C1, C2, C3, and C4 are the first-class SWA indices as shown in Table 1. Moreover, according to the evaluation index system given in Table 1, C1=XC11; C2=a·XC21+b·XC22; C3=c·XC31+d·XC32 and C4=e·XC41+f·XC4 2, where the parameters (a–f) can be determined through expert judgment and XC11,XC21,XC22,XC31,XC32,XC41, and XC41 are the standard values of the second-class soil water availability indices (C11,C21,C22,C31,C32,C41,C42). It should be noted that the values of the second indices must be transformed to be standard values according to the requirements of the AHP method. There are two types of second-class indices. One is the positive index, which means that when the second index is larger, the first-class index is larger. The other is the negative index, which means that when the second-class index is larger, the first-class index is lower. Among the second-class indices in this study, the indices C21 and C22 are the negative indices, which are transformed to be standard values by the following method:

$$XC = \frac{X_{max} - C}{X_{max} - X_{min}} \tag{14}$$

where XC represents the index standard value, which ranges from 0 to 1; C represents the actual calculated value of the second-class index; Xmax represents the theoretical maximum value of the second-class index; and Xmin represents the theoretical minimum value of the second-class index.

The remaining five indices (C11, C31, C32, C41, and C42) are the positive indexes, which are transformed to be standard values by the following method:

$$XC = \frac{C - X_{min}}{X_{max} - X_{min}} \tag{15}$$

The terms are all described in Equation (14). Based on the above analysis, the comprehensive SWA is calculated by the following equation:

$$U = 0.07XC_{11} + 0.13XC_{21} + 0.06XC_{22} + 0.12(XC_{31} + XC_{32}) + 0.1XC_{41} + 0.4XC_{42} \tag{16}$$

where the terms are the same as those described previously.

2.3. Overview of the Study Area

Handan City is located in the southern part of the North China Plain at latitude 36.07° N to 37.02° N and longitude 113.47° E to 115.47° E, which covers an area of 12,047 km^2 (Figure 2). Handan City has a semi-arid continental monsoon climate with four distinct seasons and high spatially and temporally distributed precipitation. According to the statistics, the average annual precipitation is 548.9 mm, and the average annual temperature ranges from 12.5 °C to 14.2 °C. According to the local Water Resources Bulletin statistics, the water consumption from households and industries in Handan City increased dramatically from 2003 to 2012. During the last decade, household and industrial water consumption increased by 6.8% annually, reaching 5.3×10^8 m^3 in 2014 and accounting for 21.8% of the total water consumption in Handan City. Moreover, the average annual runoff water resources of Handan City have decreased dramatically from 5.8×10^8 m^3 (1956–2000 runoff data series) to 4.2×10^8 m^3 (2001–2010 runoff data series). Owing to the sharp increase in water consumption and the sharp decrease in runoff water resources, agricultural water has been diverted to other water sectors, which is a great threat to the sustainable development of agriculture and food security in this region.

Figure 2. Location map of the study area.

Handan City has a long history of agricultural cultivation and is one of the most important grain and cotton production areas in the North China Plain. According to the Year Book of Agricultural Statistics, the annual grain production in Handan City was above 4.5×10^9 kg from 2005 to 2012, ranking in the top three areas in Hebei Province. Winter wheat, cotton, and summer maize are the dominant crops. The average farmland area in this region in the past decade reached 660,000 ha, accounting for almost 10% of the total arable land in Hebei Province. For these reasons, Handan City was selected as the study area.

One of the most important objectives of this research is to simulate the soil moisture in the farmland. In MODCYCLE, the Green–Ampt method is used to simulate the soil water infiltration and percolation process using land use data, plant variety, and soil type as key parameters. In this research, the land use data in the study area were retrieved by remote sensing. Five main land use patterns are included in the study area: farmland, forest land, grassland, urban areas, and rural residential areas. Farmland covers the largest area of Handan City, accounting for 86% of the total area. The crop variety data in this research were collected from the Rural Statistical Yearbook of Hebei Province combined with the field survey. For the field survey, we visited more than 100 farmers to determine the crop varieties and planting areas, particularly in Daming and Quzhou County. The survey data were used to verify the reliability of the Rural Statistical Yearbook of Hebei Province. The statistics indicate nine crop varieties in the study area, with winter wheat, multiple cropping of wheat and summer maize, cotton, vegetables, and apple orchard as the five dominant crops. The soil-type data were collected from the China Soil Scientific Database and were validated through high-density sampling experiments. Generally, 22 soil types are included in the study region. The distribution maps of the planting varieties with land use and soil type in the study area are separately shown in Figure 3a,b.

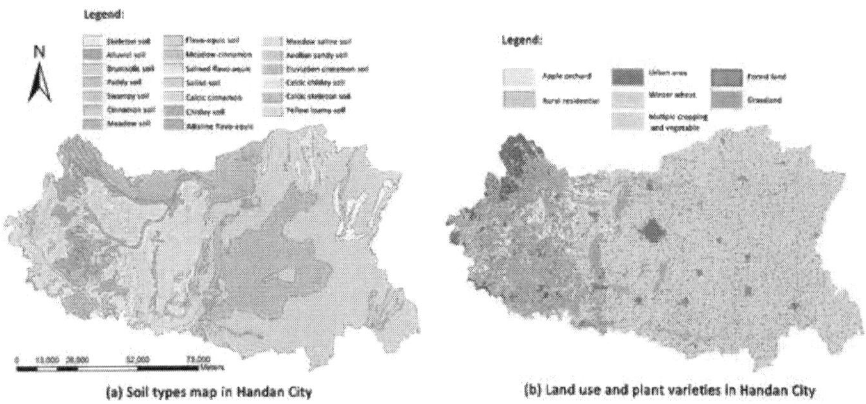

Figure 3. Distribution of soil and land use in Handan City.

3. RESULTS

3.1. Model Construction

The input parameters required in the model construction process include basic geographic information, observed meteorological data, agricultural management regimes, and hydrological data. The basic geographic information includes topographical data, land use and crop varieties, soil type, and the main digital rivers and reservoirs in the study area, which are all used for delineation of the sub-basins and discretization of the HRUs. The observed meteorological data include precipitation, temperature, wind speed, radiation, and relative humidity at a daily timescale. The agricultural management regimes are important in the model because they depict the planting configuration and irrigation scheduling in the farmland. The hydrological data include river runoff, inflow and outflow of the reservoirs, and the observed soil moisture, which can be used for model calibration and validation. For convenience, only the observed soil moisture data were used for the initial calibration of the model. The data required for MODCYCLE are shown in Table 3.

Table 3. Data required for MODCYCLE model construction.

Data Types	Detail	Source
Basic geographic information	1. DEM (90 m × 90 m)	Land Processes Distributed Active Archive Center of USA [41].
	2. Land use map (1:100,000)	The map was retrieved by remote sensing data, which are provided by Data Center for Resources and Environmental Sciences, Chinese Academy of Sciences (RESDC): http://www.resdc.cn
	3. Soil property map (1:100,000)	China Soil Scientific Database: http://www.soil.csdb.cn/
	4. Digital river and reservoir (1:250,000)	The digital river and reservoir map was drawn based on the paper map using GIS tool
	5. Groundwater aquifer data	The data were collected from Handan Water Conservancy Bureau
Meteorological observed data	1. Daily rainfall 2. Average daily temperature 3. Average daily wind speed 4. Daily sun radiation 5. Average relative humidity	China Meteorological Data Network [42].
Agricultural management regimes	1. Crop varieties 2. Planting configuration 3. Irrigation regimes	The data were collected from "Rural Statistical Yearbook of Hebei Province"
Hydrological data	1. Runoff in 5 main rivers 2. Inflow and outflow in 12 reservoirs 3. Soil moisture data measured in 12 monitoring stations	The data were collected from Handan Water Conservancy Bureau
Others	1. socio-economic water consumption data.	The data were derived from "Statistical Yearbook of Handan"

In this study, the period of the simulation was 1998–2014. A warm-up period from 1996 to 1997 was also used for this model. All of the data were collected within this period. The data from 1998 to 2000 were used for model calibration, and those from 2001 to 2003 were used for model validation. Model calibration is an important process for improving the simulation accuracy and precision. Ten parameters were selected as the most sensitive parameters. Because the MODCYCLE is a newly developed model, the auto calibration module is not yet embedded. Therefore, we adopted the manual parameter adjustment method for parameter calibration. After more than 20 attempts and simulations, the optimal set of parameters were determined and selected, as shown in Table 4.

Table 4. Key parameters and their optimal values used in MODCYCLE.

NO.	Parameter	Description	Optimal Value
1	MXSP	Maximum water depth stored in the surface of farmland, which is closely related to runoff generation	100 mm
2	ESCO	Correction coefficient in calculating soil evaporation from the soil profile, which is an important factor affecting water volume evaporated in different soil layers	0.70
3	EPCO	Correction coefficient in calculating the plant evapotranspiration from the soil profile, which affects water volume absorbed by plant in different soil layers	0.96
4	FFCB	The ratio between initial soil moisture and field capacity of the soil profile	0.6
5	GWEC	Coefficient in calculating the water volume recharged from groundwater aquifer to soil profile	0.89
6	SLAG	Runoff lag coefficient	3.2
7	SOLK	The hydraulic conductivity of different soil layers	2–150 mm/h
8	SOLA	Depth of available water in the soil profile	100–600 mm
9	CANM	Maximum water depth intercepted by crop canopy	6 mm
10	BIOE	The radiation-use efficiency of the plant	15–40 (kg/ha)/(MJ/m^2)

3.2. Model Validation

In our study, we compared the simulated and soil observed water contents to validate the model. The observed soil water content data were obtained by sampling in the soil profile, where the average depth of the soil profile was usually 600 mm. Four sampling points were selected in one soil profile at depths of 0–15 mm, 15–30 mm, 30–45 mm, and 45–60 mm. After the soil samples were collected, the soil water content was measured using the oven drying method. Based on the above measurements, the average soil water content in the entire soil profile was calculated as the weighted average value of the soil water content of the four sampling points. We collected the in situ observed soil moisture data from 12 soil moisture monitoring stations evenly distributed in the study area. A list of these stations is given inTable 5. For convenience, the 12 monitoring stations were also classified into four groups.

For validation, the simulated soil water content results in the HRUs in which the soil moisture monitoring stations are located were selected for comparison with the observed soil water content from the 12 monitoring stations. The depth of the soil water content simulated in MODCYCLE was as much as 3000 mm. However, to keep the depth of the simulation results consistent with the observed results, we extracted only the simulated soil water content of the shallow soil layer (0–60 mm) to make comparisons with the observed soil moisture data. According to the simulation and observed results, comparison charts were drawn separately for the four groups of soil moisture monitoring stations, as shown in Figure 4.

Table 5. Soil moisture monitoring stations in the study area.

Group	ID	Station Name	County	Water Source	Latitude	Longitude
Group-1	1	Paihuai Station	Wu'an	Rain-fed	114.02	36.62
	2	Guantai Station	Cixian	Rain-fed	114.08	36.33
	3	Kuang-menkou Station	Shexian	Rain-fed	113.78	36.45
Group-2	4	Xin-an Station	Feixiang	Rain-fed	114.68	36.55
	5	He-hengcheng Station	Cheng'an	Rain-fed	114.57	36.48
	6	Quzhou Station	Quzhou	Rain-fed	114.97	36.78
Group-3	7	Linming-guan Station	Yongnian	Rain-fed	114.48	36.67
	8	Ping-gudian Station	Guangping	Rain-fed	115.07	36.57
	9	Cai-xiaozhuang Station	Weixian	Irrigation	114.93	36.28
Group-4	10	Long-wangmiao Station	Daming	Irrigation	115.22	36.22
	11	Linzhang Station	Linzhang	Irrigation	114.62	36.35
	12	Wei-sengzhai Station	Guantao	Irrigation	115.38	36.72

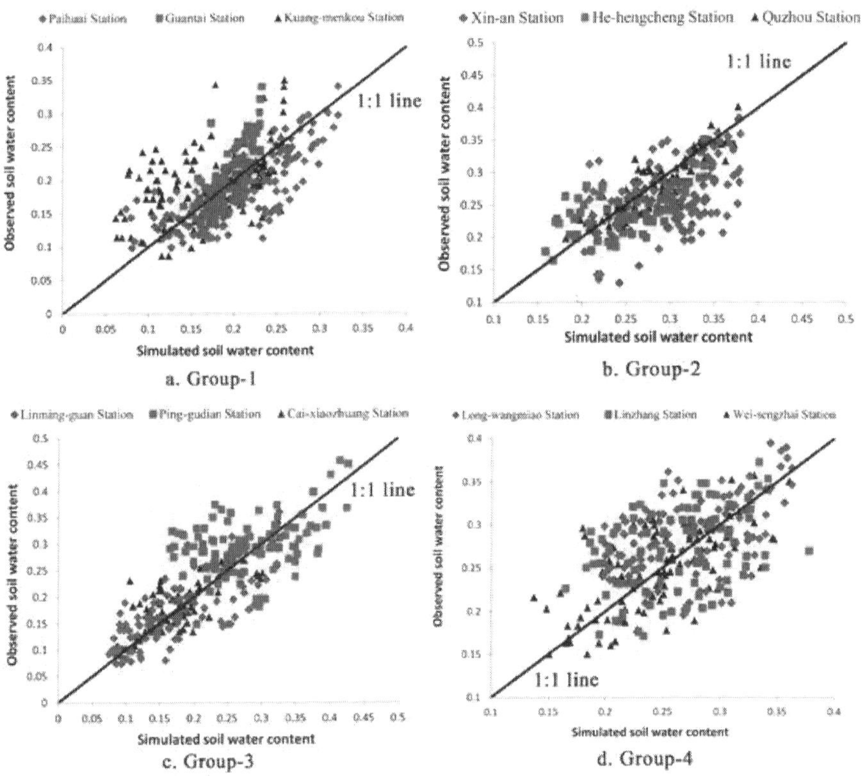

Figure 4. Comparison of observed values and simulated values of the 12 monitoring stations.

As shown in Figure 4, the points in which the x- and y-coordinates represent the simulated and observed values, respectively, were essentially distributed along the 1:1 lines. This demonstrates an obvious consistency between the simulated and observed results. To further illustrate and validate the simulation results, the relative error (ER) and correlation coefficient (R) were introduced to quantitatively depict the consistency between the simulated and the measured results. As shown inTable 6, the statistical results show good agreement between the simulated and observed soil water contents as well.

Table 6. Statistical results showing the consistency between observed and simulated values.

ID	Station Name	Observed Average Value	Simulated Average Value	ER (%)	R
1	Paihuai Station	0.173	0.189	8.83	0.87
2	Guantai Station	0.201	0.187	6.94	0.82
3	Kuang-menkou Station	0.155	0.152	1.79	0.95
4	Xin-an Station	0.275	0.301	9.33	0.76
5	He-hengcheng Station	0.241	0.253	4.75	0.72
6	Quzhou Station	0.241	0.253	4.75	0.75
7	Linming-guan Station	0.248	0.243	2.02	0.95
8	Ping-gudian Station	0.261	0.264	1.15	0.97
9	Cai-xiaozhuang Station	0.195	0.187	3.93	0.72
10	Long-wangmiao Station	0.313	0.289	7.45	0.6
11	Linzhang Station	0.257	0.26	1.47	0.52
12	Wei-sengzhai Station	0.248	0.255	2.82	0.77

As shown in the table, the relative errors (ER) were all below 10%, and the correlation coefficients (R) were all above 0.7 except for the data of Long-wangmiao and Linzhang stations. Because these two stations are located in the east and south plain area of Handan City, where the anthropogenic influence on farmland hydrological process is stronger than that in other areas, the simulation accuracy is poorer compared with other stations. This problem is also a global concern in hydrology. Generally, the simulation results were convincible and acceptable. Therefore, it is feasible to use MODCYCLE for soil water availability evaluation in the study area.

3.3. Soil Water Availability Quantitatively Evaluation

3.3.1. Temporal Perspective

As previously mentioned, the simulation period in this study was from 2001 to 2014. Based on the simulation results, the comprehensive soil water availability in the entire study area from 2001 to 2014 was calculated, as shown in Figure 5.

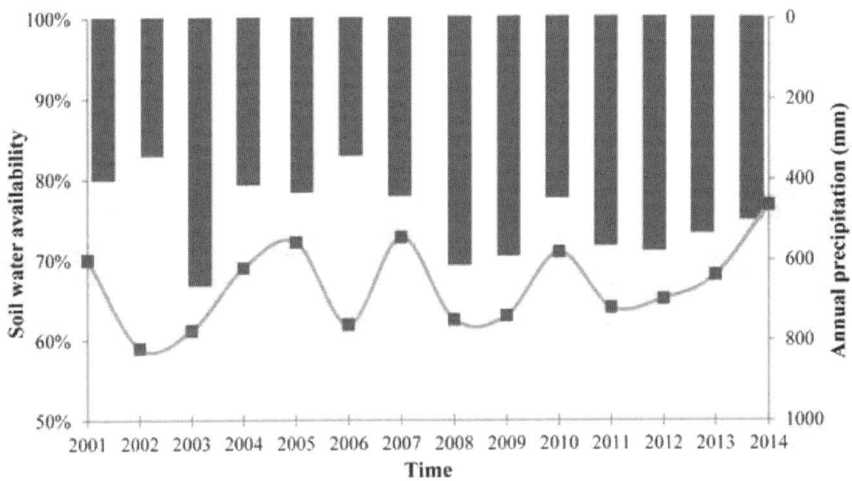

Figure 5. Comprehensive soil water availability in the simulation period.

Figure 5 shows that the comprehensive SWA ranged from 59% to 76.8% in different years. The maximum value of 76.8% occurred in 2014 with yearly precipitation of 507 mm, and the minimum value of 59% occurred in 2002 with yearly precipitation of 346 mm. To analyse the relationship between soil water availability and the yearly precipitation of different hydrological level years, we ranked the SWA from 2001 to 2014 in ascending order. In addition, we determined the hydrological level of the different years according to the frequency analysis of precipitation. The resultant SWA ranking chart is shown inFigure 6.

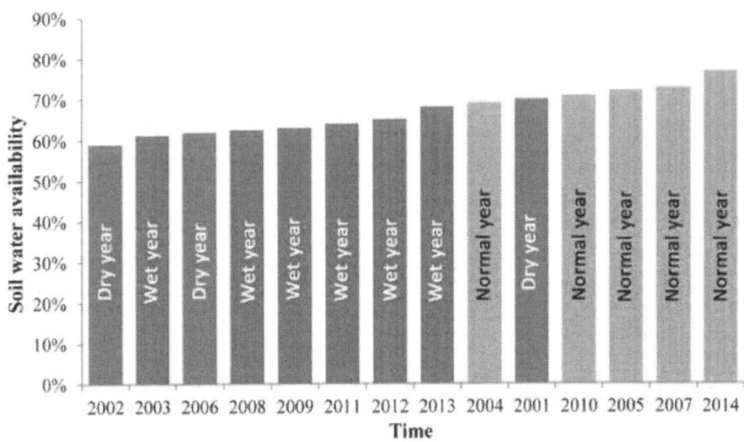

Figure 6. SWA ranking in different hydrological level years.

Figure 6 shows the SWA from 2001 to 2014 sorted in ascending order. The high-soil-water-availability years were concentrated mainly in normal hydrological years. Five normal hydrological years were indicated in the simulation period, in which the SWA, shown by yellow bars in the figure, was much higher than that in dry and wet years, shown by brown and blue bars, respectively. Preliminary analysis indicated that in dry years, the water supply from precipitation was deficient, which cannot satisfy plant water requirements during the growing period. Therefore, the poor degree of matching between the soil water supply and plant water demand is the primary reason for the low SWA in dry years. In wet years, the precipitation was concentrated mostly in summer. Therefore, the available soil water distribution was extremely uneven, which in turn resulted in significant water loss in the rainy season, leaving plants vulnerable to drought in the sowing season.

3.3.2. Spatial Perspective

The spatial distribution of the comprehensive SWA was mapped based on the simulation results. As the first step, the original values of the second-class indices were quantitatively calculated on the basis of Equation (3) to Equation (13). In this study, three different typical hydrological-level years were selected (e.g., a wet year, normal year, and dry year) through the frequency analysis of precipitation. The years 2006, 2010, and 2012 were determined to be typical dry, normal, and wet years. As the second step, the original values of the second-class indices were transformed to the standard values as shown inTable 7, Table 8 and Table 9 according to Equations (14) and (15). As the final step, the comprehensive SWA in different administrative regions was calculated on the basis of Equation (16).

As shown in Table 7, Table 8 and Table 9, because the index C11 represents the valid capacity of the soil reservoir, which is largely determined by the soil properties in the root zone, the standard values of the index in the three different hydrological years are the same. Spatial distribution maps of SWA in normal, wet, and dry years, based on the standard values of the second-class indexes, are shown inFigure 7, Figure 8 and Figure 9, respectively.

Table 7. Standard values of the second-class indices and the total SWA of different regions in the normal year (2010).

Normal Year (2010)	XC11	XC21	XC22	XC31	XC32	XC41	XC42	U
Wu'an Region	0.416	0.836	0.806	1.00	0.82	0.915	0.513	0.701
Jize Region	0.899	0.730	0.754	0.83	0.55	0.900	0.611	0.703
Qiuxian Region	0.983	0.377	0.864	0.93	0.96	0.943	0.525	0.701
Yongnian Region	0.848	0.736	0.710	0.83	0.63	0.859	0.606	0.701
Quzhou Region	0.981	0.617	0.836	0.83	0.56	0.910	0.579	0.688
Handan Region	0.723	0.680	0.839	0.95	0.58	0.913	0.625	0.714
Feixiang Region	0.914	0.572	0.840	0.83	0.51	0.931	0.689	0.718
Guantao Region	0.895	0.709	0.806	0.83	0.52	0.917	0.659	0.720
Shexian Region	0.100	0.976	0.723	0.94	0.90	0.872	0.698	0.764
Guangping Region	0.921	0.629	0.840	0.83	0.51	0.922	0.686	0.724
Cheng'an Region	0.919	0.493	0.844	0.83	0.53	0.933	0.685	0.710
Weixian Region	0.879	0.629	0.813	0.83	0.49	0.932	0.691	0.720
Cixian Region	0.573	0.785	0.800	0.83	0.71	0.877	0.511	0.667
Linzhang Region	0.893	0.517	0.842	0.83	0.53	0.922	0.704	0.717
Daming Region	0.771	0.741	0.734	0.94	1.00	0.921	0.668	0.786
Urban Region	0.281	0.889	0.871	0.95	0.75	0.891	0.529	0.692
Fengqu Region	0.355	0.896	0.702	1.00	0.61	0.869	0.541	0.680
Average in the whole area	0.727	0.672	0.798	0.87	0.66	0.910	0.623	0.712

Table 8. Standard values of the second-class indices and the total SWA of different regions in the wet year (2012).

Wet Year (2012)	XC11	XC21	XC22	XC31	XC32	XC41	XC42	U
Wu'an Region	0.416	0.784	0.608	0.89	0.84	0.873	0.414	0.628
Jize Region	0.899	0.728	0.684	0.77	0.57	0.849	0.551	0.665
Qiuxian Region	0.983	0.358	0.874	0.81	1.00	0.921	0.529	0.689
Yongnian Region	0.848	0.729	0.711	0.77	0.52	0.839	0.560	0.659
Quzhou Region	0.981	0.641	0.774	0.77	0.63	0.894	0.568	0.683
Handan Region	0.723	0.686	0.703	0.77	0.52	0.865	0.542	0.640
Feixiang Region	0.914	0.727	0.671	0.77	0.54	0.857	0.562	0.666
Guantao Region	0.895	0.728	0.663	0.77	0.57	0.892	0.618	0.694
Shexian Region	0.100	0.975	0.491	0.9	0.72	0.708	0.526	0.639
Guangping Region	0.921	0.704	0.601	0.77	0.55	0.856	0.581	0.668
Cheng'an Region	0.919	0.737	0.652	0.77	0.54	0.866	0.531	0.655
Weixian Region	0.879	0.704	0.543	0.77	0.57	0.842	0.566	0.657
Cixian Region	0.573	0.757	0.66	0.87	0.5	0.741	0.394	0.574
Linzhang Region	0.893	0.682	0.673	0.77	0.55	0.845	0.572	0.663
Daming Region	0.771	0.778	0.435	0.89	1.00	0.770	0.501	0.685
Urban Region	0.281	0.882	0.765	0.87	0.76	0.807	0.392	0.613
Fengqu Region	0.355	0.881	0.534	0.77	0.53	0.724	0.396	0.558
Average in the whole area	0.727	0.716	0.64	0.81	0.65	0.838	0.525	0.649

Table 9. Standard values of the second-class indices and the total SWA of different regions in the dry year (2006).

Dry Year (2006)	XC11	XC21	XC22	XC31	XC32	XC41	XC42	U
Wu'an Region	0.416	0.659	0.737	0.82	0.45	0.924	0.468	0.591
Jize Region	0.899	0.302	0.755	0.65	0.43	0.940	0.588	0.606
Qiuxian Region	0.983	0.121	0.861	0.71	0.45	0.945	0.484	0.564
Yongnian Region	0.848	0.248	0.776	0.65	0.48	0.933	0.629	0.619
Quzhou Region	0.981	0.178	0.791	0.65	0.47	0.942	0.556	0.590
Handan Region	0.723	0.381	0.718	0.65	0.54	0.929	0.573	0.608
Feixiang Region	0.914	0.339	0.581	0.65	0.50	0.922	0.599	0.613
Guantao Region	0.895	0.338	0.752	0.65	0.46	0.937	0.613	0.624
Shexian Region	0.100	0.923	0.658	0.6	0.60	0.875	0.650	0.658
Guangping Region	0.921	0.298	0.683	0.65	0.47	0.931	0.633	0.625
Cheng'an Region	0.919	0.330	0.518	0.65	0.53	0.928	0.593	0.610
Weixian Region	0.879	0.351	0.603	0.65	0.88	0.932	0.635	0.674
Cixian Region	0.573	0.440	0.808	0.59	0.73	0.931	0.520	0.605
Linzhang Region	0.893	0.298	0.587	0.65	0.90	0.935	0.626	0.666
Daming Region	0.771	0.427	0.612	0.61	0.74	0.927	0.639	0.657
Urban Region	0.281	0.722	0.785	0.59	0.52	0.942	0.492	0.585
Fengqu Region	0.355	0.712	0.713	0.65	0.61	0.931	0.545	0.622
Average in the whole area	0.727	0.370	0.694	0.66	0.60	0.931	0.585	0.619

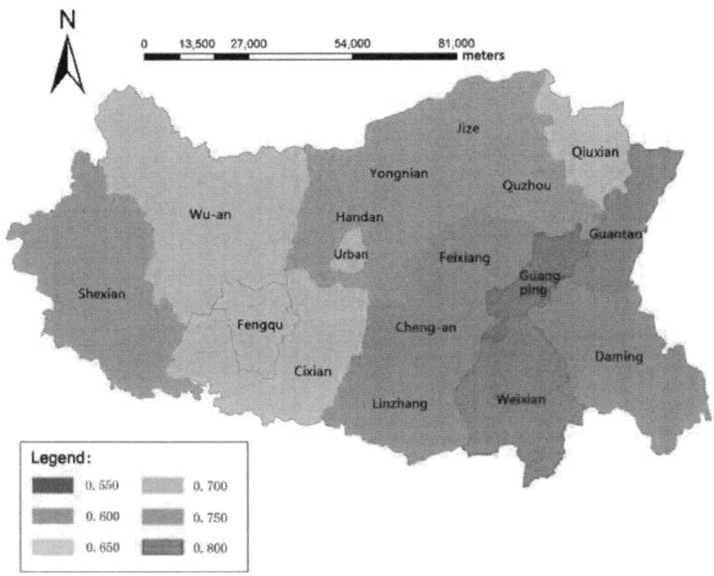

Figure 7. Spatial distribution of SWA in the normal hydrological year (2010).

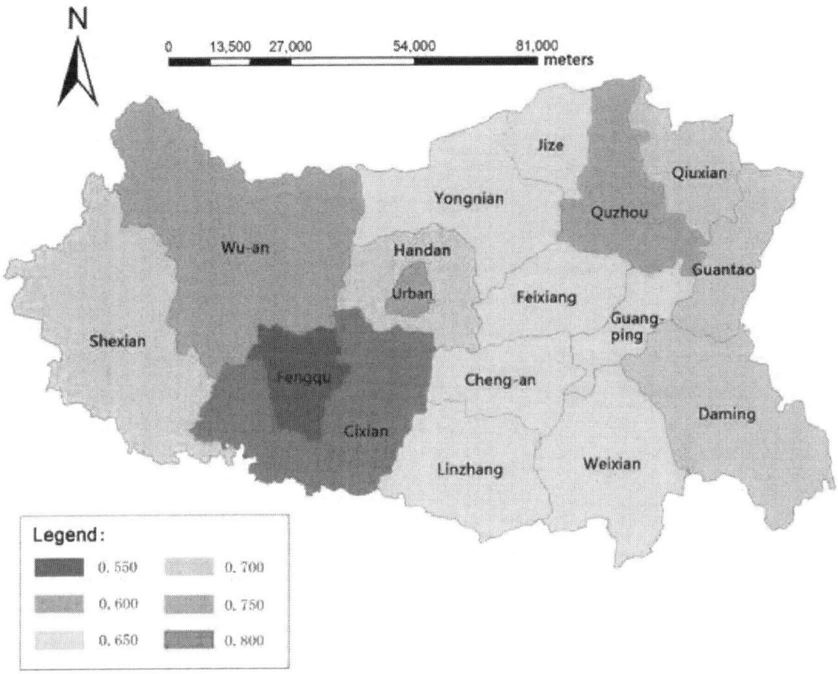

Figure 8. Spatial distribution of SWA in the hydrological wet year (2012).

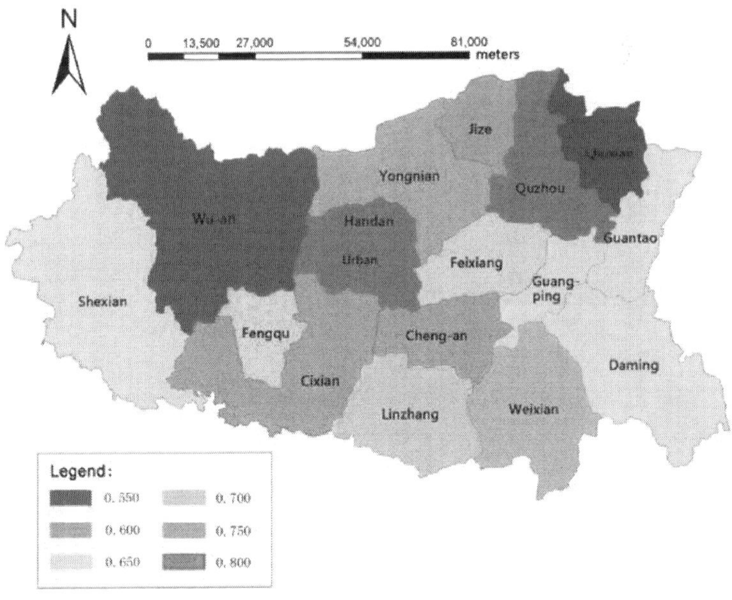

Figure 9. Spatial distribution of SWA in the hydrological dry year (2006).

According to Table 7, Table 8 and Table 9 and Figure 7, Figure 8 and Figure 9, the average SWA of Handan City is 0.712, 0.649, and 0.619 in the hydrological normal, wet, and dry years, respectively. In general, when the SWA is below 0.7, the farmland soil water use efficiency (SWUE) is too low to be acceptable; when the SWA is above 0.8, the SWUE can be regarded as satisfactory. With this guidance, the average SWA of Handan City in the wet and dry years is not acceptable. Although the SWA in the normal year is higher than in the wet and dry years, there is still great potential for improvement. From a spatial distribution perspective, the SWA in the west and north mountainous area is lower than that in the east and south plain areas. This may be attributed to the following reasons. (1) The total precipitation in the south and east areas is higher than that in the north and west mountainous areas, and the annual temporal distribution of precipitation in the south and east areas is relatively even, which will increase the degree of matching between the soil water supply and crop water demand and will therefore improve the SWA in the south and east areas; (2) The root zone depth in the west and north mountainous area is thin. Therefore, the factor of the soil water storage capacity is smaller than that of the east and south plain areas; (3) Because the quality of arable land in the eastern and southern plain areas is better than that in the mountainous areas, multiple agricultural management measures such as irrigation and cultivation are implemented in the plain areas, which will improve the use efficiency of the agricultural water resources in the farmland.

Overall, the maps show that the variation in the SWA in different regions of the study area is large and that there is still great potential for improvement in the SWA and SWUE. The SWA distribution maps provide a visual display of its spatial and temporal distribution, allowing easy detection of the problematic regions for targeting specific measures for soil water regulation.

4. DISCUSSION

4.1. Analysis of the Main Factors Influencing the Soil Water Availability for Different Regions

As shown in Figure 7, Figure 8 and Figure 9, the SWA varies greatly in different regions because the values of the influencing factors (C11, C21, C22, C31, C32, C41, and C42) differ among regions. To analyze the main factors influencing the SWA for each region, we separately calculated the contribution ratio of each factor to the SWA in each region, and we plotted the statistical radar maps as shown in Figure 10and Figure 11.

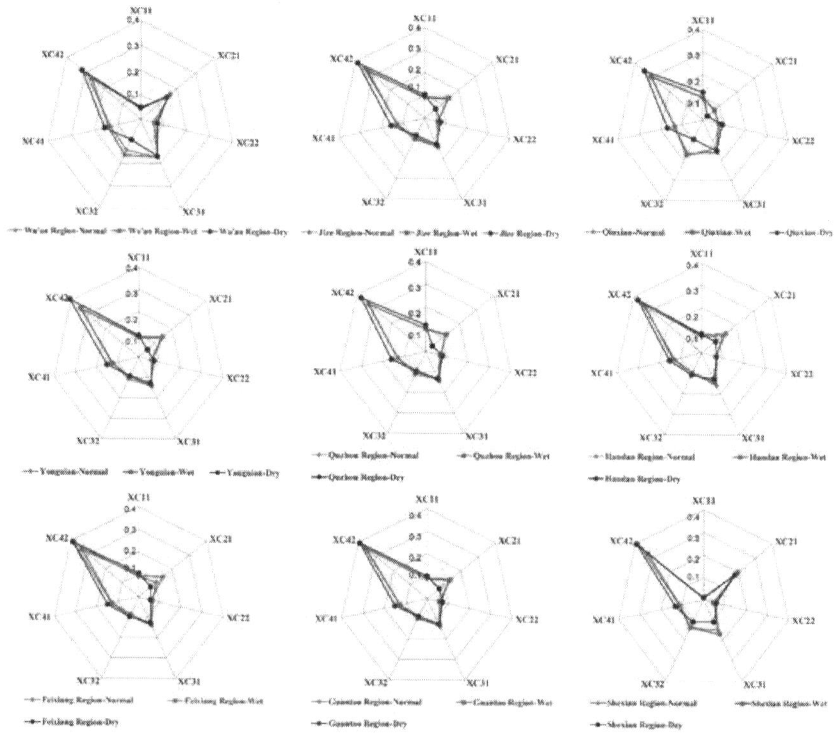

Figure 10. Contribution ratio of different factors to the comprehensive SWA in Wu'an, Jize, Qiuxian, Yongnian, Quzhou, Handan, Feixiang, Guantao, and Shexian regions in different hydrological-level years. As shown in Table 1, the factor of C11 represents the valid physical capacity of the soil reservoir, C21 represents the average value of the annual empty storage capacity of the soil reservoir, C22 is the factor for the average annual variation of the empty storage capacity of the soil reservoir, C31 represents the factor for the temporal matching degree between the soil water supply and crop water demand, C32 represents the factor for the spatial matching degree between the soil water supply and crop water demand, C41 is the factor for the soil water conversion efficiency, and C42 is the factor for the crop use efficiency of the soil water.

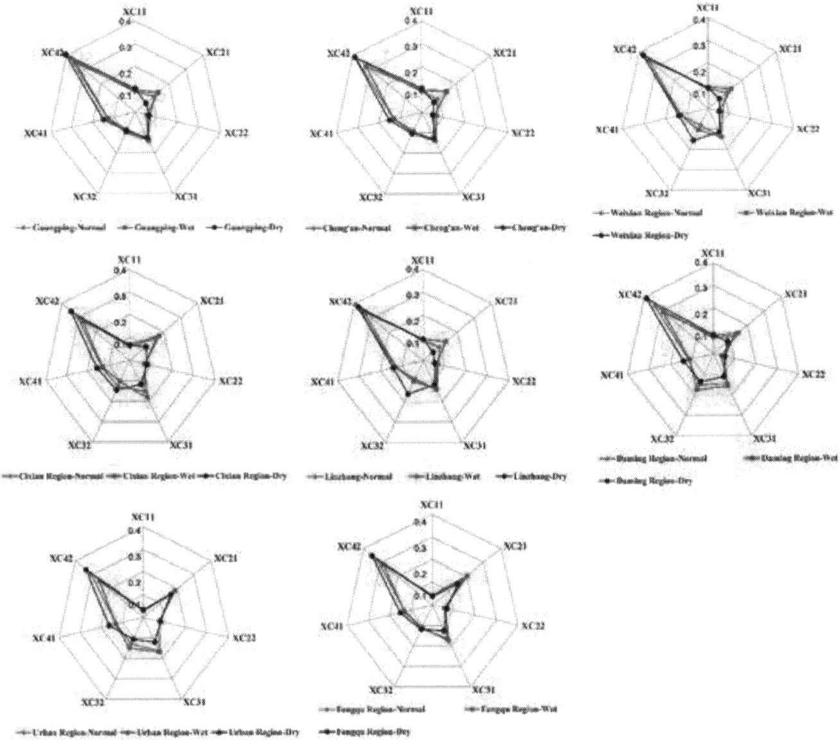

Figure 11. Contribution ratio of different factors to the comprehensive SWA in Guangping, Cheng'an, Weixian, Cixian, Linzhang, Daming, Fengqu, and urban regions in different hydrological-level years. As shown in Table 1, the factor of C11 represents the valid physical capacity of the soil reservoir, C21 represents the average value of the annual empty storage capacity of the soil reservoir, C22 is the factor for the average annual variation of the empty storage capacity of the soil reservoir, C31 represents the factor for the temporal matching degree between the soil water supply and crop water demand, C32 represents the factor for the spatial matching degree between the soil water supply and crop water demand, C41 is the factor for the soil water conversion efficiency, and C42 is the factor for the crop use efficiency of the soil water.

According to the radar maps, the contribution rate of different factors to the SWA in different regions can be clearly illustrated. Among the seven main factors, C42 is the most important with a contribution rate above 0.3 except for Wu'an, Cixian, and urban regions. The factors C21, C31, and C41 are also important with contribution rates between 0.1 and 0.2. Comparatively, C11, C22, and C32 contribute less to the total SWA with contribution rates around or below 0.1. According to the above analysis, to improve the total SWA, we should focus on the four factors (C42, C21, C31, and C41) and take measures to target increases in the contribution rates of these four factors. To analyse the potential for improvement in these contribution rates, we defined an index K to indicate the difference between the current value of the factor and the desired value of the factor, as shown in Equation (17):

$$K_n = (V_d - V_c) / V_d 100\% \tag{17}$$

where Kn is the potential of the factor n to be improved to a higher contribution rate to SWA, and Vdis the maximum value of the contribution rate of the factor n, which is determined on the basis of Equation (16). The coefficient of each factor in Equation (16) can be regarded as the maximum value of the contribution rate of the corresponding factor; Vc is the current value of the contribution rate of the factor n.

The potential of each factor in normal hydrological years for different regions was calculated on the basis of Equation (17), as shown in Table 10.

Table 10. Potentials of the factors to be improved for a higher contribution to SWA in normal hydrological years.

Region	K_{C11}	K_{C21}	K_{C22}	K_{C31}	K_{C32}	K_{C41}	K_{C42}
Wu'an Region	58.4%	16.4%	19.4%	0.0%	18.0%	8.5%	48.7%
Jize Region	10.1%	27.0%	24.6%	17.0%	45.0%	10.0%	38.9%
Qiuxian Region	1.7%	62.3%	13.6%	7.0%	4.0%	5.7%	47.5%
Yongnian Region	15.2%	26.4%	29.0%	17.0%	37.0%	14.1%	39.4%
Quzhou Region	1.9%	38.3%	16.4%	17.0%	44.0%	9.0%	42.1%
Handan Region	27.7%	32.0%	16.1%	5.0%	42.0%	8.7%	37.5%
Feixiang Region	8.6%	42.8%	16.0%	17.0%	49.0%	6.9%	31.1%
Guantao Region	10.5%	29.1%	19.4%	17.0%	48.0%	8.3%	34.1%
Shexian Region	90.0%	2.4%	27.7%	6.0%	10.0%	12.8%	30.2%
Guangping Region	7.9%	37.1%	16.0%	17.0%	49.0%	7.8%	31.4%
Cheng'an Region	8.1%	50.7%	15.6%	17.0%	47.0%	6.7%	31.5%
Weixian Region	12.1%	37.1%	18.7%	17.0%	51.0%	6.8%	30.9%
Cixian Region	42.7%	21.5%	20.0%	17.0%	29.0%	12.3%	48.9%
Linzhang Region	10.7%	48.3%	15.8%	17.0%	47.0%	7.8%	29.6%
Daming Region	22.9%	25.9%	26.6%	6.0%	0.0%	7.9%	33.2%
Handan Region	71.9%	11.1%	12.9%	5.0%	25.0%	10.9%	47.1%
Fengqu Region	64.5%	10.4%	29.8%	0.0%	39.0%	13.1%	45.9%

According to the analysis in Table 10, it is concluded that the potential for improvement of the factors differs among regions. For example, in the Wu'an Region, the indices KC11 and KC42 were much higher those of other factors, which means that the factors C11 and C42 have a large potential to be increased for a higher contribution to SWA. However, in the Jize Region, the indices KC32 and KC42 were much higher than those of other factors, which means that the factors C32 and C42 have large potential to be increased for a higher contribution to SWA. The values of the index K changed dramatically among the regions. Therefore, to improve the SWA in the study area, different measures must be taken according to the different potential of the factors in different regions.

4.2. Strategies to Improve Soil Water Availability

The concept of SWA is a new comprehensive indicator to quantitatively characterise the SWUE of farmland at a regional scale. This indicator not only considers the traditional index of field water use efficiency but also considers the soil water storage capacity of the soil layer, the actual capacity of the soil reservoir, and the spatial and temporal matching degree between the soil water supply and crop water demand. Therefore, SWA is a better index for SWUE evaluation and soil water management.

According to the analysis in Table 10, to improve the SWA, it is necessary to take different measures in different regions because the potential of the factors to be improved for a higher contribution to SWA is different.

First, in the Wu'an, Shexian, Cixian, Handan, and Fengqu regions, the factor C11 has a large potential for improvement because the index K is above 40% in the five regions and is as high as 90% in the Shexian Region. Therefore, using agronomic measures such as deep ploughing and straw returning to increase the valid capacity of the soil layer in the five regions is a key step for improving the SWA.

Second, in the Jize, Quzhou, Handan, Feixiang, Guantao, Guangping, Cheng'an, Weixian, and Linzhang regions, the factor C32 has a large potential for improvement because the index K is above 40% in the nine regions. This indicates that the degree of spatial matching between the soil water supply and crop water demand is very poor in the nine regions. Therefore, the planting structure in these regions should be adjusted to reduce the area of water-consuming crops such as winter wheat or to enlarge the area of water-efficient crops such as summer maize and potatoes.

Finally, as shown in Table 10, KC42 was mostly above 30% in all 17 regions, although the index was only 29.6% in the Linzhang Region. Obviously, the potential of improving the C42 factor is quite large in all of the regions in the entire study area. Therefore, increasing the soil water use efficiency for crop growth, such as mulching to reduce invalid evaporation and adopting regulated deficit irrigation, is a key approach for increasing the SWA.

5. CONCLUSIONS

Under the joint effects of climate change and human activities, the availability of blue water resources is severely reduced for agricultural production, particularly in arid and semi-arid areas. Therefore, soil water, known as green water resources, is becoming more and more important for sustainable development in agriculture and ecology. In this context, to improve the SWUE, this study put forward a new method to quantitatively evaluate the SWA based on a self-developed hydrological model at the regional scale and presented related strategies according to the main factors leading to low efficiency of soil water utilization. In summary, the main innovations in this research include two aspects. Firstly, this study originally proposed the concept of SWA and established a new evaluation index system to quantitatively calculate the SWA through the analytic hierarchy process method. Secondly, a self-developed distributed hydrological model (MODCYCLE) was introduced to simulate the soil moisture change process for SWA evaluation. However, modelling farmland soil moisture through a hydrological model is still challenging because of the high variability of underlying surface conditions and intensive human management measures, particularly in the Chinese farming system. Therefore, this study attempted the novel development of MODCYCLE considering the complicated tilling practices and management schemes in the North China Plain. Different from the SWAT model, MODCYCLE includes a special agricultural management module to depict the complex hydrological processes of Chinese farmland. Therefore, the simulation accuracy is greatly improved by using MODECYCLE to evaluate the SWA in the study area with complicated tilling practices and strong anthropogenic influences.

To verify the SWA evaluation method and the MODCYCLE model, Handan City in North China Plain was selected as a case study. The results show that the MODCYCLE model is an efficient tool for simulating the farmland hydrological process in the study area. According to the comparison between the observed soil moisture and simulated soil moisture results, the relative errors are all below 10%, and the correlation coefficients are almost above 0.7 except for Long-wangmiao and Linzhang stations. The SWA evaluation results indicate that the average SWA in the entire area ranges from 0.62 to 0.71 in different hydrological level years, which demonstrates that the potential is large for SWA improvement for a higher SWUE. However, the soil water availability changes greatly in different regions because the main factors influencing the SWA are different. Based on these findings, we suggest that managers can use the SWA index system to identify the main factors causing low soil water utilization efficiency in different regions and to select and determine the strategies for improving the SWA. Improving the reliability of the hydrological model is the foundation for SWA evaluation. However, owing to the complicated tillage practices and irrigation schemes in China, the farmland hydrological characteristics vary extensively among different regions, which poses a great challenge for model calibration and validation. Therefore, more attention should be paid to model improvement and validation when the model is adapted to other regions in future studies.

ACKNOWLEDGMENTS

The researchers thank the Chinese National Natural Science Foundation (No. 51409275, No. 51309246 and No. 71503265) for their support of this study. The study was also supported by Natural Science Basic Research Plan in Shaanxi Province of China (No. 2016JQ5076), the Open Research Fund of the State Key Laboratory of Simulation and Regulation of Water Cycle in River Basin at the China Institute of Water Resources and Hydropower Research (IWHR-SKL-201601) and the Ph.D. Start-up Fund of Northwest A&F University (2452015344). The help provided by Wenxiang Pan, Miao Sun, and Qi Zhao is also appreciated.

AUTHOR CONTRIBUTIONS

Xuerui Gao established the soil water availability evaluation system and constructed the MODCYCLE model. Jianhua Wang proposed the concept of soil water availability and designed the preliminary road map for this study. Pute Wu and Yong Zhao analysed the data and validated the model. Xining Zhao and Fan He revised the first draft.

CONFLICTS OF INTEREST

The authors declare no conflict of interest.

REFERENCES

1. Xu, X.; Yang, D.; Yang, H.; Lei, H. Attribution analysis based on the Budyko hypothesis for detecting the dominant cause of runoff decline in Haihe basin. *J. Hydrol.* **2014**, *510*, 530–540.
2. Lema, M.A.; Majule, A.E. Impacts of climate change, variability and adaptation countermeasures on agriculture in semiarid areas of Tanzania: The case of Manyoni District in Singida Region, Tanzania. *Afr. J. Environ. Sci. Technol.* **2009**, *3*, 206–218.
3. Robinson, D.A.; Campbell, C.S.; Hopmans, J.W.; Hornbuckle, B.K.; Jones, S.B.; Knight, R.; Ogden, F.; Selker, J.; Wendroth, O. Soil moisture measurements for ecological and hydrological watershed scale observatories: A review. *Vadose Zone J.* **2008**, *7*, 358–389.
4. Walker, J.P.; Houser, P.R.; Willgoose, G.R. Active microwave remote sensing for soil moisture measurement: A field evaluation using ERS-2. *Hydrol. Process.* **2004**, *18*, 1975–1997.
5. Deans, J.D.; Milne, R. An electrical recording soil moisture tensiometer. *Plant Soil* **1978**, *50*, 509–513.
6. Kodikara, J.; Rajeev, P.; Chan, D.; Gallage, C. Soil moisture monitoring at the field scale using neutron probe. *Can. Geotech. J.* **2014**, *51*, 332–345.
7. Dasberg, S.; Dalton, F.N. Time domain reflectometry field measurements of soil water content and electrical conductivity. *Soil Sci. Soc. Am. J.* **1985**, *49*, 293–297.
8. Dirksen, C.; Dasberg, S. Improved calibration of time domain reflectometry soil water content measurements. *Soil Sci. Soc. Am. J.* **1993**, *57*, 660–667.
9. Tanaka, D.L.; Aase, J.K. Fallow method influences on soil water and precipitation storage efficiency. *Soil Till. Res.* **1987**, *9*, 307–316.
10. Zegelin, S.J.; White, I.; Jenkins, D.R. Improved field probes for soil water content and electrical conductivity measurement using time domain reflectometry. *Water Resour. Res.* **1989**, *25*, 2367–2376.
11. Gao, X.; Wu, P.; Zhao, X.; Wang, J.; Shi, Y.; Zhang, B.; Tian, L.; Li, H. Estimation of spatial soil moisture averages in a large gully of the Loess Plateau of China through statistical and modeling solutions. *J. Hydrol.* **2013**, *486*, 466–478.

12. Wagner, W.; Lemoine, G.; Rott, H. A Method for Estimating Soil Moisture from ERS Scatterometer and Soil Data. *Remote Sens. Environ.* **1999**, *70*, 191–207.

13. Schmugge, T.J.; Kustas, W.P.; Ritchie, J.C.; Jackson, T.J.; Rango, A. Remote sensing in hydrology.*Adv. Water Resour.* **2002**, *25*, 1367–1385.

14. Vereecken, H.; Huisman, J.A.; Bogena, H.; Vanderborght, J.; Vrugt, J.A.; Hopmans, J.W. On the value of soil moisture measurements in vadose zone hydrology: A review. *Water Resour. Res.* **2008**,*44*, W00D06.

15. Ulaby, F.T.; Dubois, P.C.; Van Zyl, J. Radar mapping of surface soil moisture.*J. Hydrol.* **1996**, *184*, 57–84.

16. Fernández-Prieto, D.; van Oevelen, P.; Su, Z. Advances in Earth observation for water cycle science. *Hydrol. Earth Syst. Sci.* **2012**, *16*, 543–549.

17. Kong, X.; Dorling, S.R. Near-surface soil moisture retrieval from ASAR Wide Swath imagery using a Principal Component Analysis. *Int. J. Remote Sens.* **2008**, *29*, 2925–2942.

18. Ford, T.W.; Harris, E.; Quiring, S.M. Estimating root zone soil moisture using near-surface observations from SMOS. *Hydrol. Earth Syst. Sci.* **2014**, *18*, 139–154.

19. Martínez-Fernández, J.; González-Zamora, A.; Sánchez, N.; Gumuzzio, A.; Herrero-Jiménez, C.M. Satellite soil moisture for agricultural drought monitoring: Assessment of the SMOS derived Soil Water Deficit Index. *Remote Sens. Environ.* **2016**, *177*, 277–286.

20. Ni-Meister, W. Recent advances on soil moisture data assimilation. *Phys. Geogr.* **2008**, *29*, 19–37.

21. Montaldo, N.; Albertson, J.D. Multi-scale assimilation of surface soil moisture data for robust root zone moisture predictions. *Adv. Water Resour.* **2003**, *26*, 33–44.

22. Reichle, R.H.; Koster, R.D. Global assimilation of satellite surface soil moisture retrievals into the NASA Catchment land surface model. *Geophys. Res. Lett.* **2005**, *32*, 177–202.

23. Kim, S. Time series modeling of soil moisture dynamics on a steep mountainous hillside. *J. Hydrol.* **2016**, *536*, 37–49.

24. Gao, X.; Lu, C.; Luan, Q.; Zhang, S.; Liu, J.; Han, D. Mapping farmland-soil moisture at a regional scale using a distributed

hydrological model: Case study in the north china plain. *J. Irrig. Drain. Eng.* **2016**, *142*, 04016029.

25. Fu, B.; Yang, Z.; Wang, Y.; Zhang, P. A mathematical model of soil moisture spatial distribution on the hill slopes of the Loess Plateau. *Sci. China Earth Sci.* **2001**, *44*, 395–402.

26. Zhang, J.E.; Lu, C.Y.; Qin, D.Y.; Guo, Y.X.; Ge, H.F. Regional runoff study based on MODCYCLE distributed hydrology model. *Trans. Chin. Soc. Agric. Eng.* **2011**, *27*, 65–71

27. Arnold, J.G.; Allen, P.M.; Bernhardt, G. A comprehensive surface-ground water flow model. *J. Hydrol.* **1993**, *142*, 47–69.

28. Mapfumo, E.; Chanasyk, D.S.; Willms, W.D. Simulating daily soil water under foothills fescue grazing with the soil and water assessment tool model (Alberta, Canada). *Hydrol. Process.* **2004**,*18*, 2787–2800.

29. Li, M.X.; Ma, Z.G.; Du, J.W. Regional soil moisture simulation for Shaanxi Province using SWAT model validation and trend analysis. *Sci. China Earth Sci.* **2010**, *53*, 575–590.

30. Zhang, J.E.; Lu, C.Y.; Qin, D.Y.; Wang, R.D. MODCYCLE—An object oriented modularized hydrological model II. Application. Shuilixuebao. *J. Hydraul. Eng.* **2012**, *43*, 1287–1295.

31. Liu, J.; Wei, C.; Xie, Q.; Zhang, W. Capacities of soil water reservoirs and their better regression models by combining "merged groups PCA" in Chongqing, China. *Acta Ecol. Sin.* **2014**, *34*, 53–65.

32. Gu, S.X.; Xiong, Y.L.; Xu, X.; Wei, C.F.; Liu, G.C. Research progress on soil reservoir and utilization rainfall resources. *Southwest China J. Agric. Sci.* **2003**, *16*, 29–32.

33. Zhang, J.; Lu, C.; Qin, D.; Liu., M. Regional "four-water" transformation based on distributed hydrological model. *Adv. Water Sci.* **2011**, *22*, 595–604.

34. Stanhill, G. Water use efficiency. *Adv. Agron.* **1986**, *39*, 53–85

35. Saaty, T.L. A scaling method for priorities in hierarchical structures. *J. Math. Psychol.* **1977**, *15*, 234–281.

36. Vaidya, O.S.; Kumar, S. Analytic hierarchy process: An overview of applications. *Eur. J. Oper. Res.***2006**, *169*, 1–29.

37. Saaty, T.L. The modern science of multicriteria decision making and its practical applications: The AHP/ANP approach. *Oper. Res.* **2013**, *61*, 1101–1118.

38. Sara, J.; Stikkelman, R.M.; Herder, P.M. Assessing relative importance and mutual influence of barriers for CCS deployment of the ROAD project using AHP and DEMATEL methods. *Int. J. Greenh. Gas Control* **2015**, *41*, 336–357

39. Wind, Y.; Saaty, T.L. Marketing applications of the analytic hierarchy process. *Manag. Sci.* **1980**, *26*, 641–658.

40. Saaty, T.L. Decision making-the analytic hierarchy and network processes (AHP/ANP). *J. Syst. Sci. Syst. Eng.* **2004**, *13*, 1–35.

41. The Ministry of Economy, Trade, and Industry (METI) of Japan; The United States National Aeronautics and Space Administration (NASA). ASTER Global Digital Elevation Model (GDEM). Available online: http://www.jspacesystems.or.jp/ersdac/GDEM/E/4.html (accessed on 12 October 2015).

42. National Meteorological Information Centre of China. China Meteorological Data Network. Available online: http://data.cma.cn/data/detail/dataCode/A.0012.0001.html (accessed on 18 February 2016).

4

CHAPTER

The Effect of Some Soil Physical and Chemical Properties on Soil Aggregate Stability in Different Locations in Sulaimani and Halabja Governorate

Salahaddin Abdulqadir Aziz, Saman Mahmood Karim

Department of Soil and Water Sciences, Faculty of Agricultural Sciences, Sulaimani University, Sulaimani, Iraq

ABSTRACT

Some soil properties were studied in relation soil aggregate along a climatologically region and different crop land use in the Northeast of Sulaimani City/Kurdistan Region of Iraq. Five locations were selected along these regions ranging from semiarid to sub-humid climatologically conditions. The soil physical, chemical properties, aggregate stability and size distribution were analyzed. A mean-weight-diameter (MWD) value was determined on 5 soils, which was the sum of the percentage of soil on each sieve (6, 3, 1.5, 0.75, 0.375 and 0.125 mm). The results of the measurements could evaluated with linear correlation coefficients for the relationships between aggregate stability (MWD) and soil physical and chemical properties of the different crop land use. The correlation coefficient for the relationship between aggregate stability and organic matter was highly significant ($P < 0.01\%$) which is in agreement with the findings of [1] . Generally large aggregates (large 6, 3, 3 - 1.5 mm) were present in highest proportions in the most semiarid of the studied areas. Aggregates 0.75 - 0.125 mm were positively correlated to fine, very fine sand and silt fractions and to organic matter. Stability of aggregates showed a positive correlation with clay content and organic matter content, while the carbonate content was strongly correlated with aggregate stability. The land use history affecting soil overlaps the pattern of climatological situations and cultivated crop lands and has to be taken into account. Aggregate size distribution and stability can be used as indicator of soil conservation and productivity.

Keywords: Aggregate Stability, Organic Matter, Some Physical and Chemical Properties, Soil Consistency Limits

1. INTRODUCTION

Soil aggregation is defined as the group of soil particles in which the forces held the particles together [2] . The aggregation is formed by some processes like physical, chemical and biological forces which are mainly responsible for their stabilization [3] [4] . The formation of aggregates are affected by some factors that are involves wetting and drying, and freezing cycles, All these processes are able to bring soil particles close together that the physical and chemical forces between them hold the particles [3] . The soil aggregates stability occurs due to the forcing activating of inorganic and organic materials. The Atterberg limit tests were later adapted by two scientists, [5] , to be used as the determination of soil behavior under different conditions. For example, one method of distinguish between clay and silt is by determining the plasticity of the soil. Plasticity refers to the response of a soil to changes in moisture content (Coduto, 128). Silt is slightly plastic, while sand and gravel do not exhibit any plasticity at all and clays can be very plastic. The plasticity index is a measure of the range of moisture contents that encompass the plastic state (Coduto 130) and is the numerical difference between the liquid and plastic limits. Soil aggregate stability is widely used to measure change in soil structure due to different management options [6] [7] . The extent of aggregate disintegration by wetting depends on aggregate stability which is related to organic matter, and clay content [8] [9] .

In General, aggregate stability is commonly measured as the percentage of water aggregates with a diameter >2 mm, a parameter which is relatively easy and faster to measure compared with mean weight diameter (MWD) [10] [11] . Many authors reported, as reviewed by [12] , that the soil aggregate stability was controlled by several soil primary characteristics, such as soil texture, clay mineralogy, contents of organic matter, and calcium carbonate. On the other hand, only few studies present about the relationship between the soil aggregate stability and environmental factors such as landscapes [13] and vegetation types [14] examined.

For predicting capability of soil for production in a given sites, the evaluation of the soil aggregate stability in relation to climatology, environmental factors and soil surface coverage would be more suitable parameters which was recognized as soil physical and chemical properties, since these parameters could be determined more readily. Thus, the objective of this study was to predicate the relationships between the soil aggregate stability and the soil surface coverage factors, these relationships will predicate the conditions raising the stability of soil aggregates, which may participate to the increase of efficient soil management to increase the soil aggregate stability.

2. MATERIAL AND METHODS

2.1. Study Area and Soil Samples

The study area is located in the Northeast of the sulaimani Governorate Iraq. Five sites were selected to carry out this study (Arbat, Bakrajow, Said Sadq, Kani Panka and Anab (Figure 1). These areas were characterized by a strong climatological gradient and cultivated with different crops from South to north, especially with respect to the annual rainfall, with a maximum in the southern site, Bakrajow with more than 750 mm and a minimum to the Northern site of less than 400 mm Anab. This climatologically gradient is also reflected in the soil temperature. Distribution and stability of soil aggregates were determined by using separate wet sieving methods [15] . The percentage by weight of aggregates at each fraction and two indexes, (MWD) Mean Weight Diameter [1] .

2.2. Physical and Chemical Analysis of Soil

For the same soil samples, chemical and physical analysis was performed as follow: particle size distribution was determined after removing the organic matter with H_2O_2 30%. [16] ; pH was measured in the extract of a soil-solution 1:2.5 in water and salinity was measured in the extract of a soil-water solution 1:5 with a conductivity meter [17] . Total calcium carbonate content was measured according to [18] . Organic matter was measured by wet oxidation ($K_2Cr_2O_7$) [19] .

2.3. Soil Consistency Limits

In this experiment, the Casagrande liquid limit device will be used to determine the liquid limit of the soil sample. The liquid limit is found when 25 drops of the liquid limit device are required to close the groove cut through the sample in a distance of 1/2. The plastic limit is found when the soil sample being rolled breaks at 1/8 in diameter. The moisture content at these limits gives a quantitative measure of these limits. The following equations for the mass, liquid limit, plastic limit, moisture content, flow index and plasticity index will be used in the determination of the Atterberg limits of the soil sample, and to further classify the soil type. Increasing soil organic matter is vital in terms of improving the soil physical and mechanical properties related to conditions for root plants and to crop development in arid and semiarid regions. This laboratory study was aimed to assess effects of organic matter on soil consistency limits (liquid limit, LL; plastic limit, PL; plastic index, PI) and soil compatibility parameters in soils with different textures (clay, silty clay, silty clay loam and clay loam). The SOC, consistency limits and plasticity index (PI = LL – PL) were measured for the soil taken from the 0 - 30 cm layer.

2.4. Aggregate Stability

Wet sieving: In this method the Soil is slowly wetted by capillarity for 30 minutes. The wetted immersed in water. The sieves are slowly raised and lowered vertically in the water by mechanical method through a distance of 1.25 inch at a rate of 30 resolutions per minute for 30 minute. The sieves in the nest have an opening from 6.0 mm to 0.125 mm [sizes (6, 3, 0.75, 0.375, and 0.125) mm] Oven dry weight of the fractions of the aggregates retained on each sieve is determined for knowing the proportionate distribution of the aggregates. The fraction of soil is retained on each sieve represents the size range viz. the size of the opening of the sieve on which the fraction of soil is retained and the size of the sieves just above through which is has passed, a wet sieving apparatus (yodder type) [20] .

3. RESULTS AND DISCUSSION

3.1. Chemical Properties

3.1.1. Effect Organic Matter Content on Aggregate Stability (MWD)

More techniques were used for evaluating soil aggregate stability, like: changes in moisture characteristic, changes in permeability, and wet-sieving [21] . Wet-sieving method was the most suitable technique. The results are expressed as a mean-weight-diameter (MWD) value, which is the sum of the percentage of soil on each sieve multiplied by the mean diameter of adjacent sieves (6.3, 1.5, 0.75, 0.375 and 0.125 mm). The results of the measurements carried out on 5 soils studied were presented in Table 2, together with linear correlation coefficients for the relationships between aggregate stability (MWD) and the locations of the soil. The correlation coefficient for the relationship between aggregate stability and organic matter (Table 1, Figure 2) was highly significant ($P < 0.01\%$) which is in agreement with the findings of [1] .

3.1.2. Effect of ($CaCO_3$, Ca^{++}, Mg^{++} and Na^+) on Aggregate Stability

Figure 3 and Figure 4, show the results of the study, effectiveness of change variance in climatology sites on chemical properties of the soil and concerned to change in aggregate stability value; in general, significant correlation determination were obtained for the relationships between aggregate stability and ($CaCO_3$, Ca^{++}, Mg^{++} and Na^+). The variation of these properties will be due, in part, to changes in the organic matter levels. Therefore, it is probable that the significance of these correlations determination is as much as reflection of the organic matter effect as it is of the effect of these properties. The correlation determination between Na^+ and aggregate stability was not significant (Table 1). Statistical analysis of the data in Bakrajow soil showed that there were higher concentration of calcium salts than magnesium salts, which is less than the predicted flocculation values, which is agree with study of the others [22] , that higher concentrations of magnesium salt were required for flocculation.

Effect of cations (Ca, Mg and Na) on aggregate stability from statistical analysis data showed in (Table 1) that the increase in (Ca and Mg) value caused increasing MWD by range from a low value 1.2 and 0.27 in Anab site to a high value 1.53 and 0.33 in kani-panka site respectively, that is located in a cool climatologically region whereas the value of MWD in this sites were 0.97 to 0.80, respectively. And the percent of Na in these sites was negative effect on aggregate stability with negative significant difference with aggregate stability.

3.2. Physical Properties

Effect soil Texture on Soil Aggregate Stability

Figure 5 and Figure 6 showed the sand [23] , silt [24] and clay fractions [8] , [25] have each been cited as soil constituents who have a major influence on the stability of soil aggregates. While this is certainly the case in the studies cited above, the correlation determination calculated between aggregate stability and these three components for the soils examined here are very small ranged in sand (0.84), in silt (0.414) and in clay (0.943) and in no way approach significance (Figure 5 and Figure 6). Although the textural range of the soils used in this study was more restricted than in the studies cited above, there are nonetheless substantial variations in the amounts of sand, silt and clay. Clay percentages range from 38 to 54, silt from 38 to 41 and sand from 8 to 19. No significant correlation was obtained with sand and silt of the particle size fractions, while significant correlation was obtained with clay content. Tisdall and Oades 1982 [25] , proposed an aggregate model depending on the size of the aggregation, that binding agents vary. Aggregates 2000 mm are held together by a fine network of roots and hyphae in soils with high contents of organic matter. Kemper and Koch (1966) [26] founded that the clay and organic matter act as cementing and binding agents in the soil. Seybold and Herrik (2001) [27] work showed that increase in clay fraction significantly improved water aggregate stability.

Table 1. Statistic analysis of the parameters studies of some the chemical properties of the soil at differences locations in sulaimani and halabja governorate.

	Treatments					
Locations	$CaCO_3$% %	O.M % %	Ca^{2+}	Mg^{2+} Meq L^{-1}	Na^{+}	MWD
Bakrajow	14a	1.22 c	1.45 ab	0.35 a	0.55 a	0.53 b
Said Sadq	17a	1.35bc	1.49 ab	0.36 a	0.44 b	0.64 b
Kani Panka	18a	1.39 bc	1.53 ab	0.33 ab	0.58 a	0.80 b
Arbat	20a	1.51 b	1.40 b	0.37 a	0.40 b	0.87 ab
Anab	15a	1.7 a	1.2 c	0.27 b	0.42 b	0.97 a

Table 2. Regression analysis of liquid limit and SOC in different soil texture class.

		Treatments					
Locations	Textural classes	Permeability (cm/hr)	Shear strength kpa	Liquid limit	Plastic limit	Plasticity index	MWD
Bakrajow	Silty Clay Loam	2.1 c	6.25 ab	27.79 e	17.38 ab	10.41 d	0.53 b
Said Sadq	Clay	3.2 b	7.10 abc	31.90 d	19.14 d	12.76 c	0.64 b
Kani Panka	Clay	3.5 b	7.35 abcd	35.75 c	20.15 c	15.6 b	0.80 ab
Arbat	Clay	3.7 ab	6.75 cd	41.98 b	21.34 b	20.64 a	0.87 ab
Anab	Clay	4.4 a	7.75 d	44.84 a	23.42 a	21.42 a	0.97 a

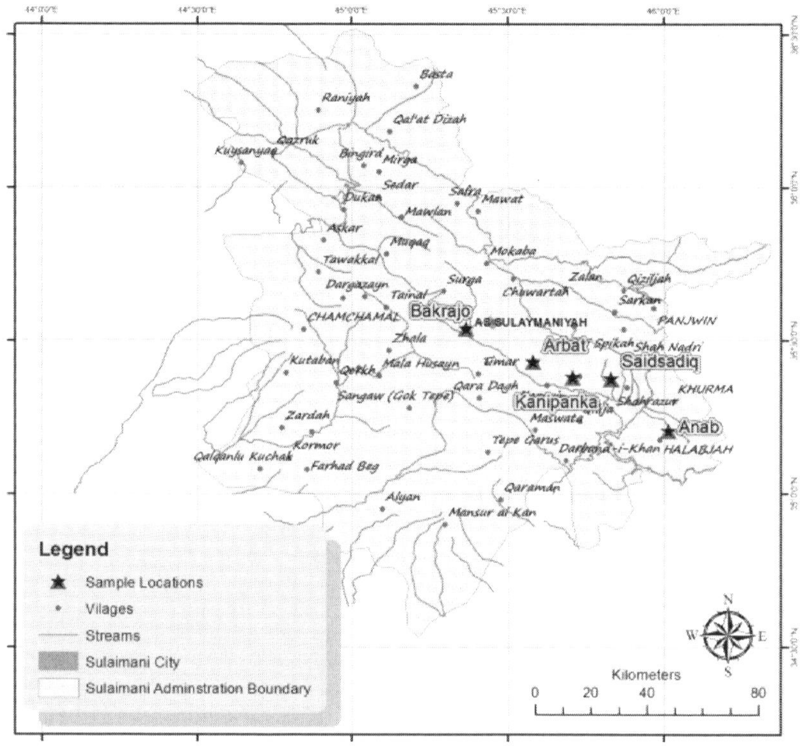

Figure 1. Some selected physical and chemical properties of the study soil at differences locations in Sulaimani and Halabja Governorate.

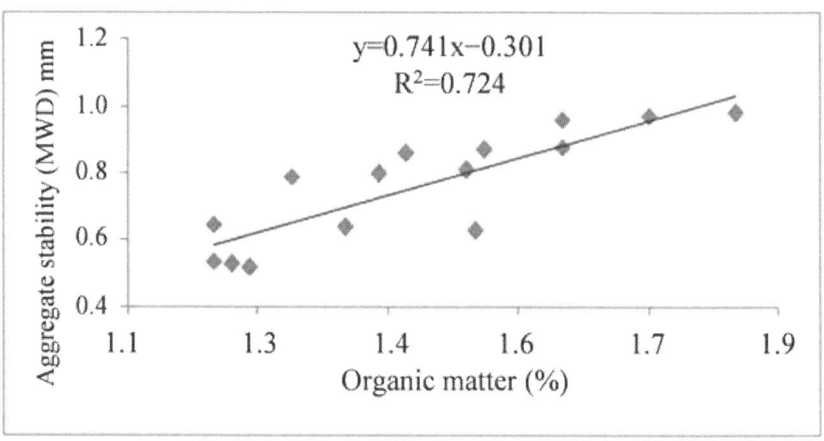

Figure 2. Relationship between aggregate stability (MWD) and organic matter (OM) content for 5 soils.

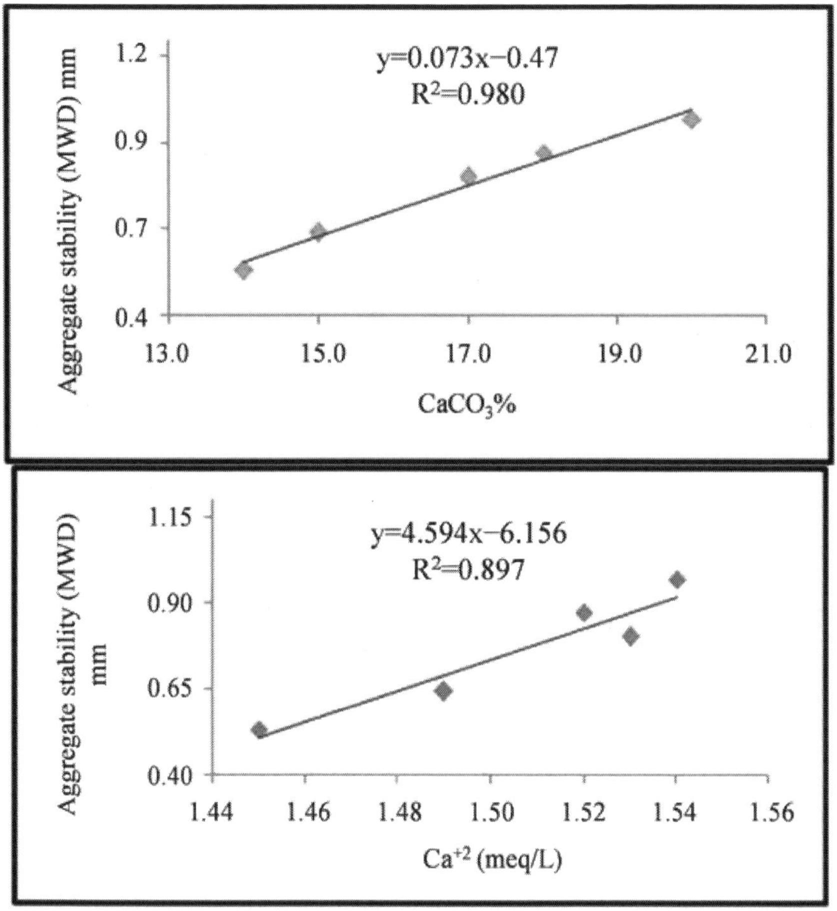

Figure 3. Relationship between aggregate stability and Calcium Carbonate, calcium content.

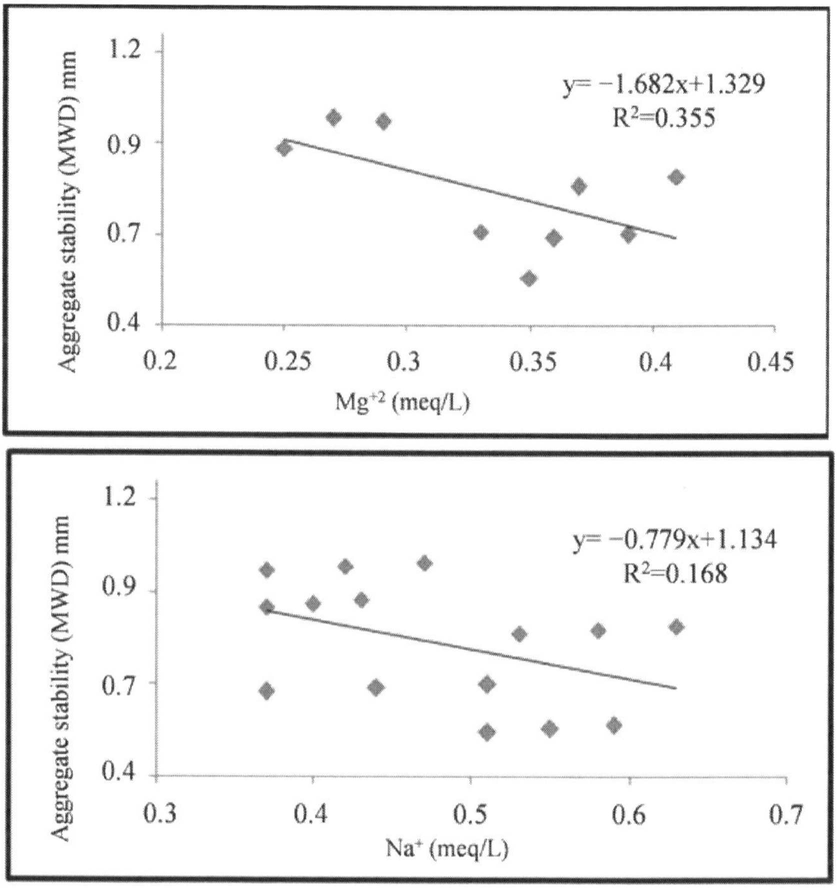

Figure 4. Influence of magnesium and sodium on aggregate stability.

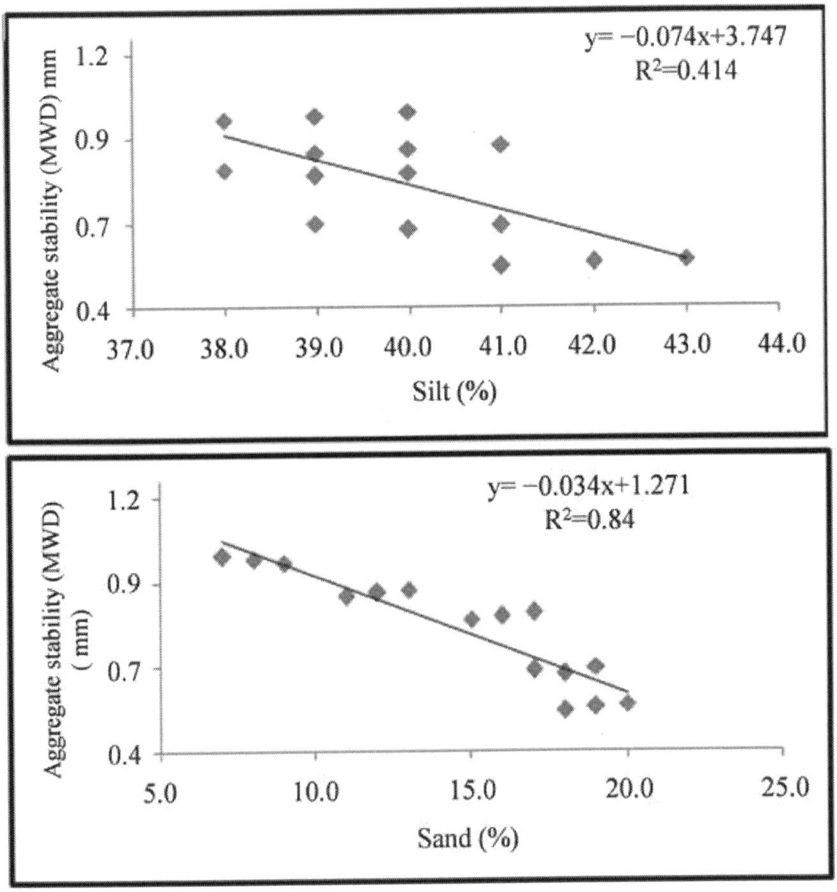

Figure 5. Corelation determination between Aggregate stability (MWD mm) and Sand and Silt %.

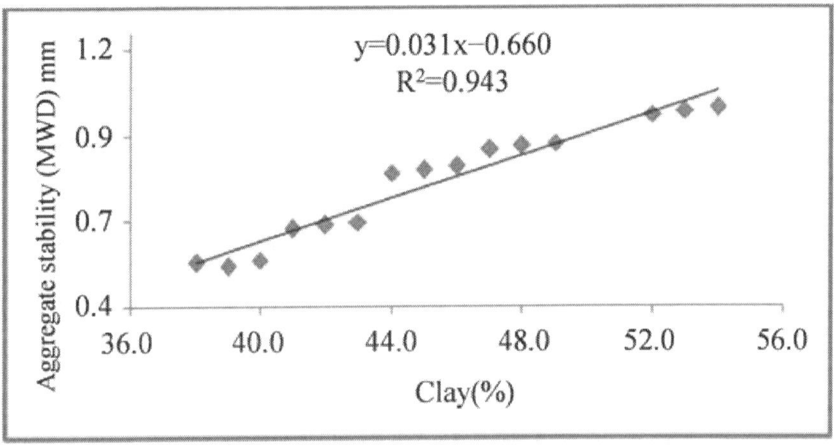

Figure 6. Correlation determination between aggregate stability (MWD mm) and Clay %.

The composition of any soil is an important factor which influences many soil properties, such as shear strength, liquid limit (LL), plastic limit (PL) permeability etc. Table 2 shows the results of this study with changes of the properties that affect on aggregate stability (MWD) from the range 6.25 to 7.75, 27.79 to 44.84, 17.38 to 23.42 and 2.1 to 4.4 cm/h as sequenced. This paper examines the effects of treatments inducing the organic matter content on the Atterberg limits of four different soils texture class. The results of regression value analysis for shear strength, liquid limit, plastic limit, and permeability in different soil texture class (Table 2 and Figures 1-4) indicated that SOC in clay soil texture class can justify 98.9 percent of the liquid limit variation. So this traits the most important component of liquid limit.

REFERENCES

1. Chaney, K. and Swift, R.S. (1984) The Influence of Organic Matter on Aggregate Stability in Some British Soils. European Journal of Soil Science, 35, 223-230.
2. Martin, J.P., Martin, W.P., Page, J.B., Raney, W.A. and De Ment, J.D. (1955) Soil Aggregation. Advances in Agronomy, 7, 1-37.
3. Allison, F.E. (1968) Soil Aggregation—Some Facts and Fallacies as Seen by a Microbiologist. Soil Science, 106, 136-143.

4. Lynch, J.M. and Bragg, E. (1985) Microorganisms and Soil Aggregate Stability. Advances in Soil Science, 2, 133-171.

5. Terzaghi, K. and Casagrande, A. (1930s) Adopted Atterberg's Tests to Civil Engineering Applications, the Liquid Limit, the Plastic Limit and the Shrinkage Limits Are Now Known as the Atterberg Limits, and Are Defined hrough ASTM's D-427 and D-4318.

6. Hussein, J., Adey, M.A. and Elwell, H.A. (1992) Irrigation and Dry Land Cultivation Effects on the Surface Properties and Erodibility of a Zimbabwe Vertisol. Soil Use and Management, 8, 97-103.

7. Salako, F.K., Babalola, O., Hauser, S. and Kang, B.T. (1999) Soil Micro Aggregate Stability under Different Fallow Management Systems and Cropping Intensities in South Western Nigeria. Geoderma, 91, 103-123.

8. Kemperw, D. and Koch, E.J. (1966) Aggregate Stability of Soils from Western United States and Canada. U. S D.A. Technical Bulletin No. 1355.

9. Kay, B.D. and Angers, D.A. (1999) Structure. In: Sumner, M.E., Ed., Handbook of Soil Science, A229-A276.

10. Elwell, H.A. (1986) Modelling Sheet Erosion and Run-Off from a Fersiallittic Clay by Combined Small-Scal Physical Simulation Techniques and Field Studies. D.Phil. Thesis, University of Zimbabwe, 233.

11. Elwell, H.A. (1987) Modeling Sheet Erosion and Runoff from a Fersiallitic Clay by Combined Small-Scale Physical Simulation Techniques and Field Studies. PhD Thesis, University of Zimbabwe, Harare, 233.

12. Le Bissonnais, Y. (1996) Soil Characteristics and Aggregate Stability. In: Agassi, M., Ed., Soil Erosion, Conservation, and Rehabilitation, Marcel Dekker, Inc., New York, 41-60.

13. Pierson, F.B. and Mulla, D.J. (1990) Aggregate Stability in the Palouse Region of Washington: Effect of Landscape Position. Soil Science Society of America Journal, 54, 1407-1412.

14. Cerda, A. (1998) Soil Aggregate Stability under Different Mediterranean Vegetation Types. Catena, 32, 73-86.

15. Dane, J.H. and Topp, G.C. (Eds.) (2002) Methods of Soil Analysis, Part 4, Physical Methods. Soil Science Society of America Book Series, No. 5, Soil Science Society of America, Madison, 1692 p.

16. Ministerio de Agricultura (1986) Métodos oficiales de análisis de suelos, III. Ministerio de Agricultura, Madrid, 166 p.

17. Richards, L.A. (1954) Diagnosis and Improvement of Saline and Alkali Soils. Agriculture Handbook, Vol. 60, USDA, Washington DC, 160 p.

18. Jackson, M.L. (1958) Soil Chemical Analysis. Prentice Hall, Englewood Cliff, 480 p.

19. Walkley, A. and Black, I.A. (1934) An Examination of the Degtjaredd Method for Determining Soil Organic Matter and Proposed Modification of the Chromic Titration Method. Soil Science, 37, 29-38.

20. Yoder, R.E. (1936) A Direct Method of Aggregate Analysis of Soils and a Study of the Physical Nature of Soil Erosion Losses. American Society of Agronomy Journal, 28, 337-351.

21. Chaney, K. (1978) The Influence of Soil Organic Matter on Soil Structure. PhD Thesis, Edinburgh University, Edinburgh.

22. Rengasamy, P., Greene, R.S.B. and Ford, G.W. (1986) Influence of Magnesium on Aggregate Stability in Sodic Red-Brown Earths. Australian Journal of Soil Research, 24, 229-237.

23. Willamrs, J.B. (1970) Relationships between the Composition of Soils and Physical Measurements Soil Aggregation as Influenced by Microbial Made on Them. Rothamsted Experimental Station Report, Part 2, 5-35.

24. Thomassona, H. (1978) Towards an Objective Classification of Soil Structure. Journal of Soil Science, 29, 38-46.

25. Tisdall, J.M. and Oades, J.M. (1982) Organic Matter and Water-Stable Aggregates in Soils. Journal of Soil Science, 33, 141-163.

26. Kemper, W.D. and Koch, E.J. (1966) Aggregate Stability of Soils from Western USA and Canada. USDA Technical Bulletin No. 1355, US Government Printing Office, Washington DC.

27. Seybold, C.A. and Herrick, J.E. (2001) Aggregate Stability Kit for Soil Quality Assessments. Catena, 44, 37-45.

5

CHAPTER

A Multi-Satellite Approach for Water Storage Monitoring in an Arid Watershed

Dawit T. Ghebreyesus [1], Marouane Temimi [1,*], Ali Fares [2] and Haimanote K. Bayabil [2]

[1]*Institute Center for Water and Environment (iWATER), Masdar Institute of Science and Technology, Masdar City, P.O. Box 54224, Abu Dhabi, UAE*

[2]*College of Agriculture and Human Sciences, Prairie View A&M University, P.O. Box 519, MS 2008 Prairie View, TX 77446, USA*

ABSTRACT

The objective of this study was to use satellite imagery to monitor the water budget of Al Ain region in the United Arab Emirates (UAE). Inflows and outflows were estimated and the trend of water storage variation in the study area was examined from 2005 to 2014. Evapotranspiration was estimated using the simplified Penman-Monteith equation. Landsat images were used to determine the extent of agricultural and green areas. Time series of gravity recovery and climate experiment (GRACE) observations over the study area were used to assess the inferred water storage variation from satellite data. The change of storage inferred from the Water Budget Equation showed a decreasing trend at an average rate of 2.57 Mm^3 annually. Moreover, GRACE readings showed a decreasing trend at a rate of 0.35 cm of water depth annually. Mann-Kendal, a non-parametric trend test, proved the presence of significant negative trends in both time series at a 5% significance level. A two-month lag resulted in a better agreement (R^2 = 0.55) between the change in water storage and GRACE anomalies within the study area. These results suggest that water storage in the study area is being depleted significantly. Moreover, the potential of remote sensing in water resource management, especially in remote and arid areas, was demonstrated.

Keywords: GRACE; Water Budget; Landsat; evapotranspiration; UAE; water resources

1. INTRODUCTION

The United Arab Emirates (UAE) has an arid climate with low average annual precipitation around 100 mm. Despite water scarcity, water consumption in the country and the region is among the highest worldwide. The disparity between available water resources and the growing water demand implies that there should be a close monitoring of water resources in the region. Freshwater in the UAE comes from groundwater and unconventional sources such as desalination. Treated water is also used in landscaping. Groundwater is mostly used for agricultural purposes and accounts for almost three-quarters of the consumption, whereas potable water is provided from the desalination plants [1].

The development of appropriate water resources management policies in the UAE requires a strong understanding of available resources and their spatial and temporal variability with respect to the growing water demand. This allows policymakers to determine effective and sustainable planning and management strategies. However, existing records on water resources are usually collected during short field campaigns and are not sufficient to develop a clear idea of the trend of water variables. The sporadic behavior of hydrological processes in arid regions and the lack of gauged stations make the monitoring of water resources challenging.

In the UAE, most of the watersheds are ungauged with very limited data records. Most of the wadis are ephemeral, especially in the western part of the country that is at the edge of the Empty Quarter. Rainfall records go back to the 1970s at only a few stations. The rest of the stations in the country are more recent. As a result, there is limited information on the hydrological behavior of the watersheds and the variability of water resources. To overcome this lack of data, the use of satellite imagery could be a potential option [2,3]. Remote sensing techniques can play a great role to fill in data gaps in ungauged watersheds, where in situ data is scarce, with their capability to provide spatial and temporal information on water resources [4,5]. This is particularly important in the case of intermittent variables like rainfall, where the common areal interpolation methods of ground records are not efficient due to the lack of dense networks of observation points. Global Precipitation products are often blended products that involve ground measurements, weather models, and satellite reading [6]. Despite the usefulness of satellite imagery for the retrieval of precipitation, the lack of ground observations impacts the quality of such products and limits their verification in arid regions [7].

In addition, water storage in reservoirs can be estimated using high resolution satellite data such as Landsat imagery and radar sensors [8,9,10,11]. The latter has been used successfully in the case of cloud presence. Storage fluctuation in reservoirs can be estimated from pre-established rating curves of volume-surface area relationships. Moreover, remote sensing techniques are used to measure changes in total water storage by quantifying the change in the total mass. Gravity recovery and climate experiment (GRACE) products were used to investigate the temporal and spatial changes of the terrestrial water resource [12,13]. This technique is more effective in areas where there is no other type of mass displaced by natural or anthropogenic mechanisms.

Recent achievements in technological advancement in remote sensing instruments are providing accurate high-resolution spatial data for key hydrological variables, especially for ungauged watersheds [14,15]. A number of hydraulic parameters such as water level height, width, and area can be captured remotely and used to estimate river discharge. In addition, river runoff, watershed concentration time, and flood wave propagation time were estimated directly from satellites with microwave sensors [9,10,16,17,18,19,20]. Nonetheless, most satellite products are at coarse spatial resolution, which is not suitable for studying hydrological processes like flooding and water storage monitoring.

In arid watersheds, like the one studied here, the scarcity of in situ data records fosters the use of satellite data at spatial and temporal resolution that would make it possible to capture the variability of water resources. To this goal the use of multi-satellite data to estimate key hydrological variables should be an attractive alternative.

In the UAE, several attempts were carried out to study the variability of water resources using remote sensing technologies. Sherif et al. [21] modeled flash floods in adjacent areas in Northern UAE. In their study, the authors used land cover maps to estimate the Curve Number (CN) across the watershed. Brook et al. [22] reported that most of the groundwater in the Al Ain region originates from eastern mountainous areas in Oman where orographic processes and microclimates trigger local precipitation events that enhance the recharge locally. The groundwater then flows westward towards the Gulf. Regionally, Milewski et al. [23] proposed a remote sensing and Geographic Information System (GIS)-based solution for estimating recharge and runoff in arid regions. Their approach was developed and tested at the regional scale. Al-Dousari et al. [24] used remote sensing data to build a Soil and Water Assessment Tool (SWAT) model for an arid watershed in Kuwait. In another study, Elmahdy and Mohamed [25] employed GIS and remote sensing data to study the variability of groundwater and its quality in the Al Ain region; but did not close the water budget and lacked a demonstration of the variability of key hydrological variables.

The city of Al Ain was selected as a case study to employ remote sensing data close the water budget. The main reason for selecting Al Ain is because the city has seen significant expansion in the agricultural sector in the last decade. Moreover, Al Ain, the third largest city in the UAE, holds a high potential of fresh groundwater in the UAE which explains the great interest in monitoring changes in water storage. The previous Water Balance studies in the area were carried out on an annual basis and, hence, they lack information on temporal evolution of water storage in the area [25,26] with limited use of satellite imagery. Additionally, the spatial coverage in previous studies was not inclusive of all the areas but rather concentrated in the city and its suburbs. Moreover, the evapotranspiration was not accurately estimated because the whole agricultural area and green areas and their expansion in time throughout the years were not accounted for.

In this study, to develop the abovementioned attempts, the potential of satellite imagery in monitoring water resources in arid regions was studied. We particularly focused on the city of Al Ain, in the UAE, and investigated the closure of the water budget using satellite data. The inland city of Al Ain is known in the UAE as the Garden City because of the relatively higher water resources potential. Close to the city, there is a lake (Lake Zakher) that receives treated wastewater which could be used as an indicator of changes in water storage in the watershed. The goal of this study is to close the water budget and examine the water storage variability in Al Ain region. The study mainly employed satellite imagery for the estimation of water storage in the lake as an indicator of changes in inflows and outflows. Water storage variability in the watershed was then compared to GRACE time series records that reflect total water storage anomalies. Although this study examined the applicability of such techniques on a small watershed scale that lies in an arid region, where most of the water inflows and outflows are controlled, results would be of paramount importance to apply similar techniques at larger watershed scales and at the country level in the UAE.

2. DATA AND METHODOLOGY

2.1. Study Area

The city of Al Ain is located in the Southeastern region of Abu Dhabi Emirate, about 130 km south of Dubai and approximately the same distance east of the capital Abu Dhabi. Its geographical coordinates are 24°12′27″ N, 55°44′41″ E. The total area of the city is approximately 13,100 km^2, with a total population of 518,316 according to 2012 census data. This area is rich in fresh groundwater compared to the other regions of the Emirate of Abu Dhabi where brackish water is dominant, especially in the western and coastal part of the country [25].

The climate in the area is hyper-arid with average annual rainfall of 100 mm [27]. Despite its arid climate, the city is attracting local and international agricultural investors due to its rich groundwater resources. The city is producing fruits and vegetables covering a significant portion of the local market demand and also features many livestock farms. Moreover, the city's industrial sector is also growing rapidly contributing to the increasing demand for water.

This watershed around the city was selected in this study primarily due to concerns that the local groundwater resources are being depleted severally in the area. This was based on recent findings that indicated the rapid depletion of water resources due to agricultural use [1]. The city of Al Ain holds a precious reserve of fresh water that is subjected to a growing demand for residential and industrial uses, as well as for agriculture given the considerable size of farms developed in the western part of the Al Ain municipality, on the road to Abu Dhabi. Therefore, understanding the water budget of the watershed is crucial to monitor the recharge mechanisms, which are very important for management and regulation purposes in a region where renewable water resources are scarce.

Lake Zakher has appeared in the desert within the study area as a result of treated wastewater discharge in 2010 (Figure 1). Since then, the lake size has been changing to indicate fluctuations in the total volume of treated water, which we believe should be in line with the change in water demand and consumption in the watershed and, therefore, a possible indicator of water storage variability. In this study this possible connection is examined.

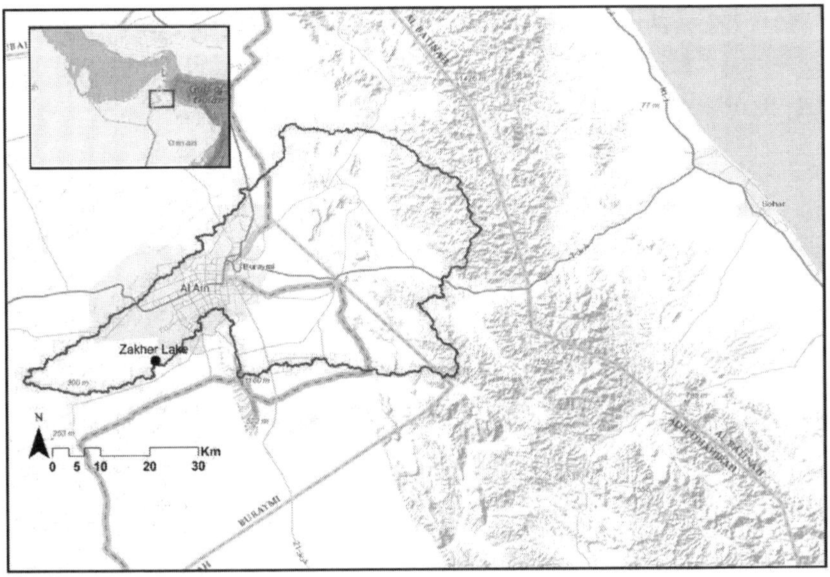

Figure 1. The city of Al Ain and its location on the Eastern United Arab Emirates (UAE) borders with Oman, and the red line shows the watershed boundary.

2.2. Dataset

2.2.1. GRACE Data

In this study, information on the total water storage variability was estimated from the Gravity recovery and climate experiment (GRACE), a mission by NASA Earth System Science Pathfinder that was launched in March 2002. GRACE consists of two identical satellites about 220 km apart and orbiting 500 km above the earth's surface. The main objective of NASA's GRACE mission is to monitor changes in the earth's mass by accurately measuring the change in distance between the satellites due to variation in gravitational force [12,28]. In this study, the level 3 gridded product from GRACE was used with 1-degree spatial resolution and approximately 1-month temporal resolution. The data was provided in anomaly relative to average values from January 2004 to December 2009, and was given in depth (cm) of water for each grid cell. The spatial resolution of the product is 100 km, which means the whole study area is enclosed in one pixel of the GRACE product. Moreover, water storage anomaly data was employed, starting from January 2005 up to December 2014 at an approximately 1-month temporal resolution.

2.2.2. Landsat Images

High resolution satellite images were used in this study to (a) delineate Lake Zhaker and monitor the change in its areal extent and (b) identify the green areas within the study domain to estimate the evapotranspiration losses. To this end, Landsat 8 imagery was used. The spatial resolution of Landsat images was 30 m with temporal resolution of 16 days. In this study, 38 cloud-free images of Landsat 7 and 8 were used covering the time period from 2010 up to 2014. Landsat images were used to capture the evolution of Lake Zhaker using the morphological gradient technique. The contour of the lake was then applied to a high-resolution 15 m Digital Elevation Model to calculate the total water storage in the lake that corresponds to each specific contour. Figure 2 shows examples of delineated lake boundary from Landsat images.

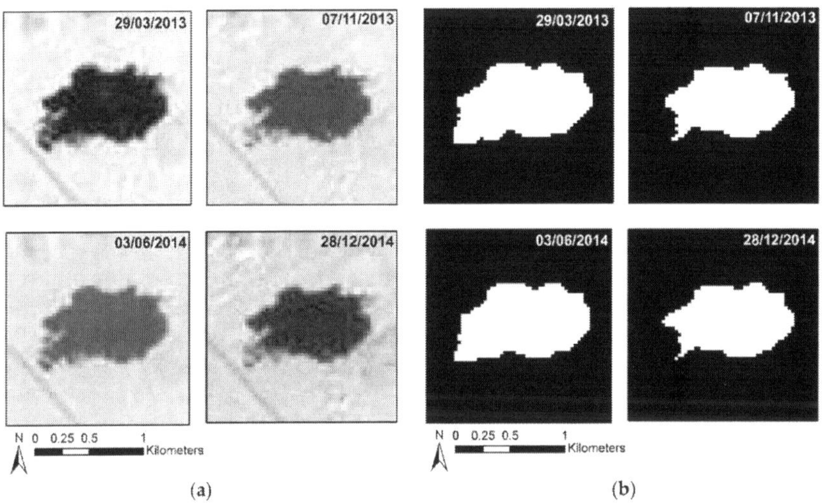

Figure 2. (**a**) Landsat 8 images at four different periods of Lake Zakher. (**b**) Extracted polygons using the morphological gradient from respective Landsat images.

2.2.3. NCEP/DOE Reanalysis II Data

We used the net incoming and net outgoing radiation data from National Center for Environmental Prediction/Department of Energy (NCEP/DOE) Reanalysis Project to determine evapotranspiration over agricultural and green areas [29]. The reanalysis information and selected outputs are available online (http://www.esrl.noaa.gov/psd/). The resolution of the data used in this study is 1.9° × 1.9°, with daily time step from 2005 up to 2013 at the surface of the earth. The resolution of the product is coarse with respect to the extent of the study area (13,000 km^2). We neglect the spatial variability of solar radiation and assumed homogenous solar radiation with the selected pixel over the study area.

2.2.4. Weather Station

In-situ observations from the weather station at the Al Ain Airport were used, starting from January 2005. Several variables were integrated into the simplified Penman-Monteith equation to estimate the evapotranspiration. The daily observations of wind speed, maximum and minimum temperature, atmospheric pressure, dew point temperature and atmospheric humidity were used on the simplified Penman-Monteith model. The precipitation data was also used as input flow in the water budget equation.

2.2.5. Water Supply, Demand, and Reuse Records

Records of water demand, consumption, recycled, and discharged water were obtained from Statistics Centre—Abu Dhabi (SCAD), which is a governmental entity responsible for the publication of official records in the Abu Dhabi Emirate [30]. Water consumption refers to the desalinated water supply, whereas recycled water denotes the treated water redistributed after treatment mostly for landscaping. However, the discharged water indicates the excess of treated water that is not reused but sent to Lake Zakher.

Records from SCAD were used along with satellite images to close the water budget in the study area. The monthly consumption of desalinated water, the annual inflow of wastewater, the annual treated water, and the annual quantity of reused water from the region of Al Ain were used.

2.3. Water Balance

A general water balance equation, $I - O = \Delta S$, where I is inflow, O is outflow, and ΔS is change in storage, was rewritten in this study as:

$$DS + P + R - Et = \Delta S \qquad (1)$$

where DS is desalinated water input to the system (watershed) , P is precipitation input to the system, R is reused water input to the system, Et is evapotranspiration from the vegetation and the lake, and ΔS is change in storage.

Inflow to the system is mainly governed by the desalinated water supply, which is common in arid regions. In Al Ain, most of the desalinated water is delivered from the desalination plant Qidfa' in the Emirate of Al Fujairah. The hybrid plant that is run by Abu Dhabi Water and Electricity Authority (ADWEA) produces 760 MW and 455,000 m^3 of water per day. Another component of the inflow to the system is precipitation, which is not as significant as the desalination water supply. The last component of the inflow is the recycled water, also known as reused water.

The main loss from the system is evapotranspiration, which includes the evaporation from Lake Zakher and the evapotranspiration from the agricultural areas within the system. Evapotranspiration was calculated using the simplified Penman equation [31] written as:

$$ET = (0.015 + 0.00042 \times T + 10 \times 10^{-6} \times z) \times (0.8 \times R - 40 + 2.5 \times F \times U \times (T - Td)) \quad (2)$$

where T is mean temperature, Td is dew temperature, U is wind speed, z is elevation, R is radiation in W/m^2, $F = (1.0 - 8.7 \times 10^{-5} \times z)$, and ET evapotranspiration mm/day.

The Penman equation combines the energy balance and mass transfer equations to estimate the evapotranspiration from an agricultural area. The modified Penman method was found to be the best approach, with minimum uncertainty compared to other methods [32]. This method has been tested in a wide range of locations and different climatic conditions. The main advantage of Penman method is its applicability in data-scarce situations. Above all, the Food and Agriculture Organization (FAO) strongly recommends the use of the Penman method where possible [33]. It is recommended to validate the outcome of the Penman method with in situ readings. Due to the absence of in situ data, the validation of evapotranspiration was not performed in this study.

The change in water storage in the study area was assessed using two approaches, the first using the determined ΔS according to Equation (1), and the second using GRACE mass anomaly values. The agreement between both approaches should corroborate the assessment of the variability of water storage variability in the area. To this end, a trend analysis was carried out using the water storage change obtained from the water balance equation and the GRACE mass change anomalies. The Mann-Kendal test was performed on both time series data at a 95% confidence level. We expected a consistency between both water storage indicators, ΔS and GRACE anomalies.

It is worth noting that in the water balance equation, the sub-surface water movement was not accounted for because of the lack of data. This will not affect the water balance model outputs as the average permeability is reported to be very small [22]. In addition, in the region and particularly in agricultural areas, deep well pumping is common as treated wastewater is not used for irrigation and desalination water that is not available in remote farms is not a cost-effective alternative. In this study, we do not account for the deep well pumping because of the lack of data. The water volumes from deep well pumping used in the irrigation could have augmented the amount of evapotranspiration. Accounting for deep well pumping would have made the estimated ΔS more accurate. However, neglecting it makes the estimation more conservative by underestimating the evapotranspiration.

3. RESULTS

3.1. Surface Water Monitoring

Zhaker Lake started to expand in late 2009, reaching 1 Mm^3 capacity in the first quarter of 2010. Then, the volume rapidly increased to reach the capacity of 4.5 Mm^3 by the year 2012. Finally, it plunged to reach a plateau of approximately 2 Mm^3 in 2013 and 2014. The increase of the lake volume in 2009 suggests the existence of an excess of recycled water that continued at least until 2012. The lake did not completely disappear after 2012; it only lost around 50% of its total volume, which may indicate that the excess of recycled water continued after 2012, perhaps at a lower magnitude.

A variability in extent can be noticed throughout the year because in 2013 and 2014 the lake size was larger in March and June than in November and December. In the region, temperature and solar radiation and, therefore, evaporation in March and June, when the lake size is larger, tend to be higher than those in November and December. This suggests that the variation in the lake size is not only sensitive to meteorological conditions but also driven by wastewater discharge.

3.2. Water Consumption

Changes in water consumption in Al Ain are shown in Figure 3. Water consumption increased with a relatively small amount, only a 30 Mm^3 increase in a four-year period between 2005 and 2009 (Figure 3). Then, water consumption started to soar in 2009 and continued, reaching 300 Mm^3 in 2013, which corresponds to an approximately 100 Mm^3 increase for the same duration (four years), a threefold increase recorded between 2005 and 2009. These records corroborate the increase in Lake Zakher volume-to-water consumption.

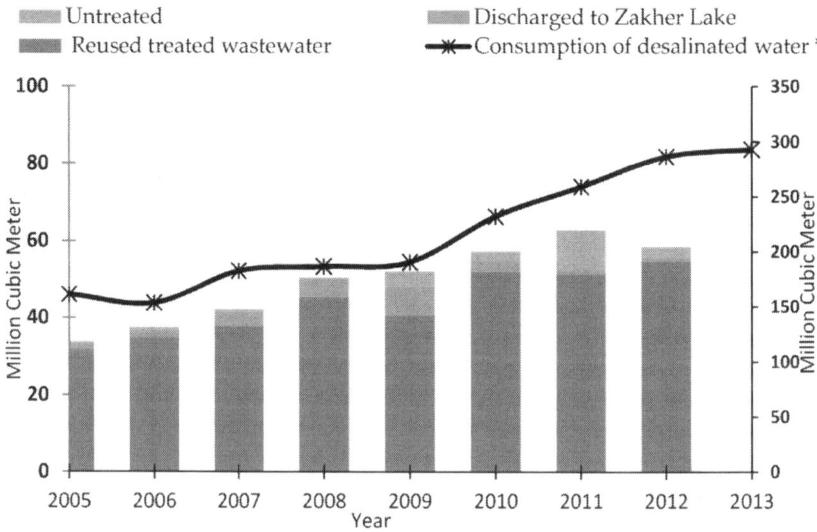

Figure 3. Desalinated water and wastewater figures from Statistics Centre—Abu Dhabi (SCAD) [30] (*Consumption of desalinated water is referred to on the right axis).

Out of 160 Mm^3 in consumed water in 2005, only 33.7 Mm^3 was sent to the wastewater treatment plants, which corresponds to 21% of the total supplied. As late as 2009, the ratio remained somehow similar, with 27% out of the total consumed reaching wastewater treatment plants, which corresponds to 52.1 Mm^3 out of the 190.9 Mm^3 in water consumed. Starting from 2009, wastewater inflow did not increase at the same rate as water consumption. Wastewater inflow increased only by 12% with the rise of more than 50% of the consumption from 2009 to 2012. This means that the increase of the wastewater inflow to the treatment plant was less than 10 Mm^3, while water consumption soared by almost 100 Mm^3. The difference between wastewater inflow and treated water was relatively small from 2005 to 2008. Increased differences were observed after 2008, with the highest difference of 10.53 Mm^3 observed in 2011.

3.3. Water Storage Monitoring

3.3.1. Water Storage from Satellite

Time series results of GRACE anomalies were used to monitor changes in water storage in the Al Ain watershed (Figure 4). Recall that we assumed that there are no other activities within the watershed involving mass movement that may impact GRACE readings, such as excessive mining or oil production activities in Al Ain watershed. It is worth noting that the spatial footprint of GRACE reading is large, with a pixel size ranging from 300–400 km. However, the main urban area, with a significant agricultural and landscaping extent within the study area, is Al Ain city, with no surface water processes elsewhere. So, we assume that anomalies in the CRACE reading should reflect fluctuations of the water storage within the city area and the agricultural areas around it. Moreover, the gridded data of level 3 (110 km × 110 km at the equator) were resampled and modified to reduce noise in the signal. The de-striping and filtering applied are essential in reducing the random measurement errors, but the methods also alter the original signal needed. This modification should affect the interpretation of the readings in the local settings (small scales) [34]. In addition, the study objective was to monitor the change in time and assess the validity of the inferred trend of storage not to quantify the actual amount of change in groundwater. So, in such contexts, discrepancies in the spatial extent should not impact the inferred trends.

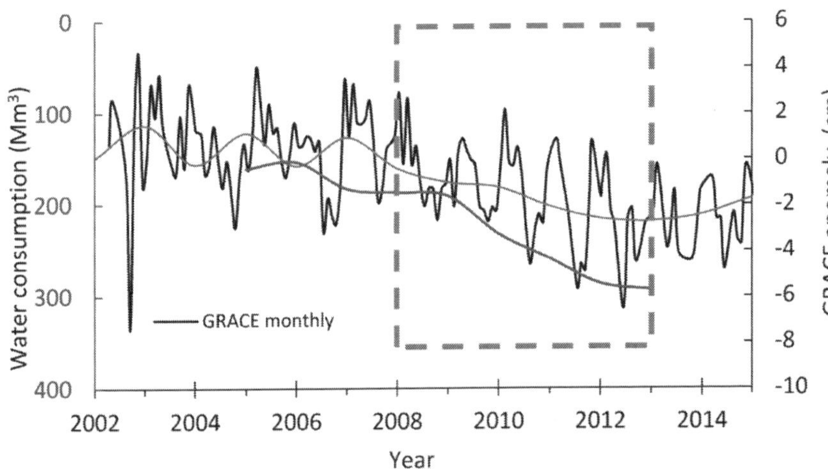

Figure 4. Grace Anomaly of Al Ain and the water consumption for the city from Abu Dhabi Water and Electricity Authority (ADWEA).

GRACE anomalies for the city of Al Ain showed a clear biannual cycle (Figure 4). Most of the positive anomalies were observed before 2008. Peak anomalies were observed during winter months of the year (December–March), while relatively low anomalies were recorded during summer months (June–October).

Water storage within the study area started to significantly decline from 2008 onwards, as the values of GRACE anomalies were persistently negative (Figure 4). The negative gradient continued and reached its lowest level (−2.7 cm) in 2013 (Figure 3). This depletion coincided with the increase in the area of Lake Zakher, which could result in the loss of a significant quantity of water to evaporation. This also indicates that groundwater resources in the area were excessively exploited in the last decade. GRACE results showed a substantial decline in the rate of depletion as the monthly and annual values flattened in 2013 and 2014. The overall rate of change of storage obtained from the GRACE readings showed a decreasing trend of 0.35 cm water depth annually. The trend analysis of the GRACE reading indicated a significant negative trend based on the Mann-Kendal test ($p < 0.05$). In agreement with observed depletion of water storage, water consumption in the city increased by 100 Mm^3 in six years (2009–2015). Water consumption reached 190.90 Mm^3 in 2009 from 161.2 Mm^3 in 2005. Then, consumption dramatically increased and reached 293 Mm^3 in 2013. The main factor for increased water demand was the expansion of the city and the growth in industrial and agricultural sectors. Figure 4shows the increase in the water consumption throughout the period of 2005–2013 versus the depletion of the groundwater resource with the GRACE results.

3.3.2. Water Balance Equation

Evapotranspiration is the main output from the system. It depends on meteorological variables, the expansion in agricultural areas and change in their extent, as well as the change in the discharged treated water and landscaping activities. In this study, the extent of agricultural area was obtained from Landsat images. Using a linear regression, the estimated annual areal increase in agricultural areas from the images was approximately 6 km^2. The estimate of the evapotranspiration of the area ranges from 50 to 350 mm/month, with an annual average of 2264 mm (based on data from 2005 to 2014). Monthly evapotranspiration data is shown in Figure 5, from 2005 to 2014. The results show the existence of two peaks of evapotranspiration over a year, in the months of April and July. The lowest evapotranspiration was recorded in the month of January, reaching as low as 50 mm/month.

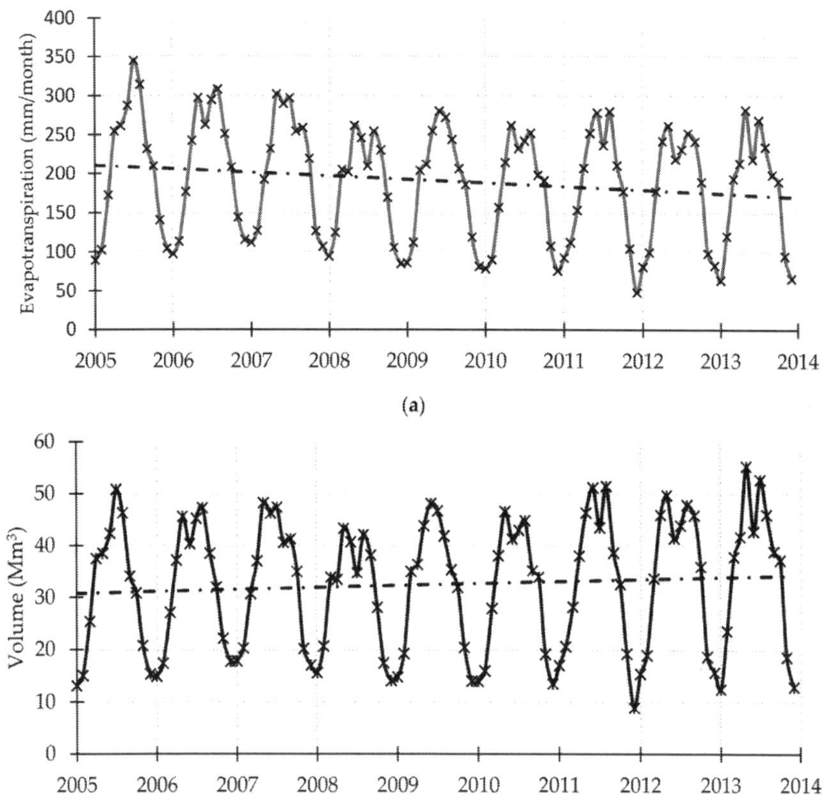

(b)

Figure 5. Time series of the monthly evapotranspiration (**a**) in mm/month over agricultural and landscaping area and (**b**) total evapotranspiration volume loss over the watershed (Mm^3/month).

Total evapotranspiration within the study area is presented in Figure 5. Unlike the monthly evapotranspiration, total evapotranspiration showed an increasing trend mainly controlled by the increase of the agricultural areas and green spaces delineated by Landsat. This suggests that expansion of irrigated areas clearly caused the increase in total evapotranspiration. As shown in Equation (1), total evapotranspiration is the major water loss from the study area. It is, therefore, expectable that the noticed increase in evapotranspiration corresponds to a depletion of water reserves in the area.

Change in water storage (ΔS) showed a strong seasonality mainly due to the variability of evapotranspiration. The rainy months (December–February) had the peak change in storage due to the relatively greater precipitation in the system compared to other months. Moreover, the evapotranspiration (loss) is low. The 2006/07 season had two extreme peaks due to unprecedented high precipitation in December 2006 (50 mm) and March 2007 (61 mm). Another noticeable increase was recorded in December 2009, with a precipitation of 74 mm/month. Generally, ΔS showed a significant negative trend at a 5% significance level using Mann-Kendal test, even though the inputs (desalinated and recycled water) showed an increase during the study period. The overall change of storage inferred from Water Budget Equation showed a decreasing trend at an average rate of 2.57 Mm^3 annually. Figure 6 represents the time series of the both the remotely sensed data and the estimated storage from the water balance equation.

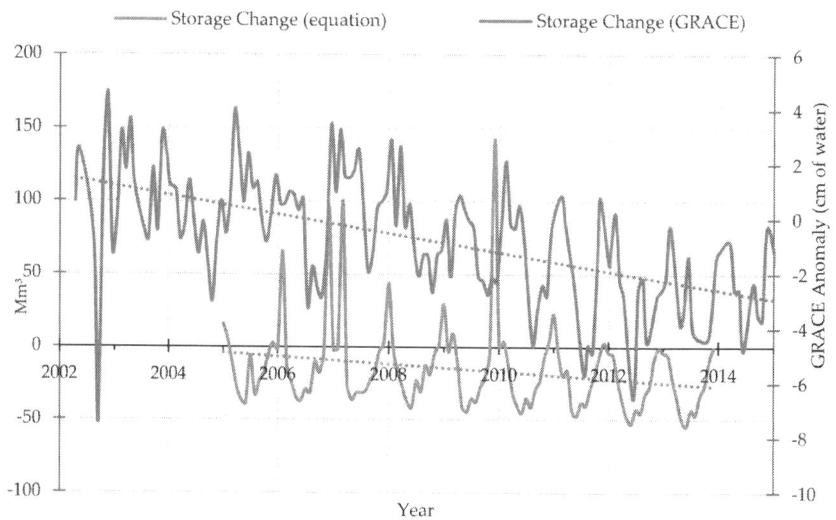

Figure 6. Variability of estimated water storage compared to gravity recovery and climate experiment (GRACE) anomalies (GRACE right axis).

3.4. Correlation between Results from Remote Sensing and Water Balance Equation

There is a clear agreement between the changes in storage calculated using the two datasets. They both show a significant negative trend when tested with the non-parametric Mann-Kendal test at a confidence level of 95%. The statistically significant negative trends in ΔS and GRACE readings corroborates the assumption of severe water depletion in the region.

A two-month lag was noticed between GRACE and water balance results. The phase lag between GRACE and water storage time series could be attributed to the difference in sampling time as GRACE data products were monthly, whereas the data used for the estimated of water storage was daily, which was then aggregated to a monthly time interval. This could also be due the fact that groundwater movement is slow to react to external sway. The correlation between GRACE anomalies and water storage estimates reaches 0.54 when a two-month lag is considered. Additionally, neglecting groundwater movement from the mountains in Oman towards the Gulf in UAE, especially during rainy periods, may have contributed to this discrepancy between both time series. However, despite the phase lag, both estimates of water storage variability from the two datasets showed significant negative trends, which corroborates the assumption of water depletion in the area.

The analysis of GRACE time series indicated two change points, 2008 and 2013, which marked a severe depletion between those specific years (Figure 7). The change point in 2008 was detected using the cumulative sum (CUMSUM) change point detection method. The CUMSUM method is a nonparametric statistical technique that can be used for detecting changes in time series [35]. The change point in 2013 was only visually identified and was not detected in the procedure mainly because of the short record after 2013. However, visually there was an existent plateau on the time series starting in 2013. The first period (2002–2008) has a gentle slope with very small slope, indicating that the water storage was relatively constant. The second period (2008–2013) indicates significant depletion of the storage followed by a period (2013–2015) of stabilization of the storage. However, this needs to be confirmed using additional years of observations.

The change points detected in Figure 7 are in line with the variability of the water consumption reported in Figure 4, where an increase in water demand coincided with the 2008–2013 time period that corresponds to a severe depletion of the water storage based on the GRACE results. Water consumption records seem to stabilize starting from the end of 2013. More recent records are needed to confirm this trend.

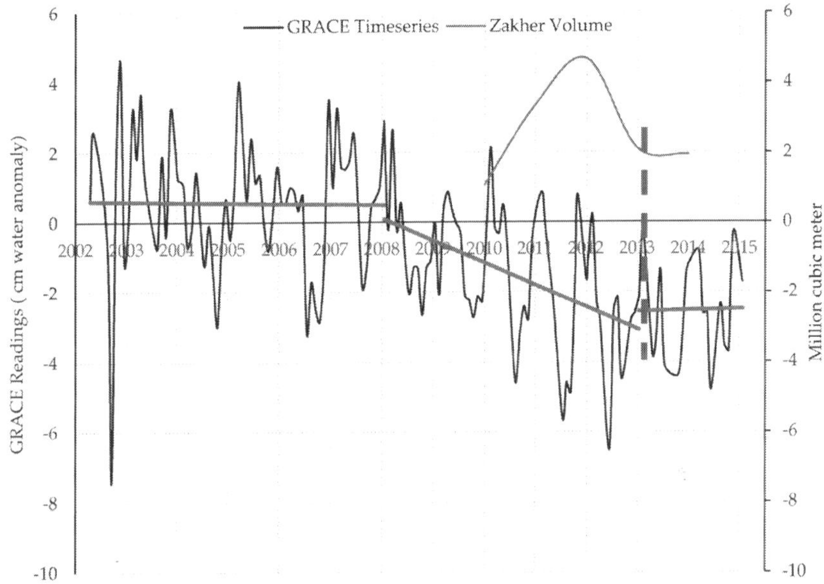

Figure 7. The Lake Zakher Volume (Mm^3) versus the GRACE time series anomalies (left axis); the dashed vertical line shows the timing of the second change point in the time series that coincides with the stabilization of the Lake Zakher water volume (Right axis).

On the other hand, the start of the depletion was followed by an increase in Lake Zakher water volume. The volume in the lake peaked in late 2011, followed by a sharp decrease in 2012 and 2013. This was in agreement with the second change point of the GRACE results. This indicates that there is a relationship between the lake's water volume and total water storage. The period of water storage increase in the lake between 2008 and 2011 corresponds to the steeper decline of water storage. The decline of water storage in the lake coincides with a slower water storage decrease between 2011 and 2013. Finally, the flattening of the water storage in the lake after 2013 corresponds to a plateau in the GRACE results. The variability of water storage in Figure 6 confirmed the depletion between 2008 and 2013. However, the record was not enough to come to a conclusion regarding the stabilization of the water depletion after 2013.

The depletion and, therefore, excess water in Lake Zakher could be also explained by the conducted infrastructure projects in the city which require draining of considerable volumes of water from the shallow water table. The drained water is usually sent to the storm water drainage network to get to the city's flood control channels or infiltrate to the domestic water network to eventually get to the wastewater treatment plant. The impact of these practices was reflected in the variability of water storage in Lake Zakher which suggests that the lake can be used as a water storage gauge that could be monitored from space.

The agricultural lands and green spaces within the study area are irrigated from desalinated and deep well pumping in the case of the former and reused water in the case of the latter. The consumption of the desalinated water is expected to increase by 22% in 2030 according to the Abu Dhabi Water and Electricity Company (ADWEC) reports. This will increase the input to the system significantly. However, the output from the system will also continue to increase if the continuous expansion of green areas (agricultural and landscaping) as reflected in the Landsat images is maintained. This suggests the need to adapt an optimized water resources management strategy in the region which would help to make informed decisions on infrastructure development. This should allow for more water reuse and advancing wastewater treatment technologies to reach a level where treated water can be used for irrigation of crops and, therefore, reduce the need for deep well water pumping.

4. CONCLUSIONS

The main goal of this study was to use satellite images to monitor water storage variability in an arid watershed. Results showed that water storage, especially groundwater, in Al Ain was depleted throughout the period of the study. This was clearly demonstrated in this study through the analysis of GRACE readings. This finding was in agreement with the previous study by [36], which indicated the increased depletion of groundwater across the UAE. Murad et al. [36] suggested that the only solution is to implement an integrated water resource management strategy. The implementation of maximum pumping regulation and controlling the expansion of the agricultural areas could be some of the initiatives to mitigate the rapid depletion of water storage in the region. Promoting other water efficient agricultural practices should contribute to the control of the water depletion in the region.

Findings from this study demonstrated that remote sensing could be effectively used for monitoring water resources in an arid watershed. Water resources can be reasonably estimated using remote sensing techniques, which can also complement in situ measurements, especially in arid and semi-arid areas where the main water input (gain) to the study area was from the desalination plant. Furthermore, the main output (loss) from the watershed was from evapotranspiration. A significant fraction of the treated water ends up in Lake Zakher, which was monitored from space to infer changes in water storage. GRACE and estimated water storage using other remotely sensed information show good agreement in showing a decreasing trend in the study area.

Overall, this study showed that remote sensing could be a useful tool in monitoring water resources in arid regions. This could be enhanced by using satellite imagery to monitor land use and vegetation health. Therefore, satellite imagery should be an important component of any comprehensive water resource management strategy that should be implemented in the UAE to achieve water and food security.

ACKNOWLEDGMENTS

This study was conducted with the support from Masdar Institute of Science and Technology grant number: SS2014-000016 entitled "Combining remote sensing and modeling techniques for estimating runoff and recharge in the arid environment of the UAE".

AUTHOR CONTRIBUTIONS

Dawit T. Ghebreyesus, an MSc student, worked on developing the results and applying the proposed methodology. Marouane Temimi, in his capacity as advisor, oversaw the application of the approach and development of the obtained results. Ali Fares and Haimanote K. Bayabil helped in editing the manuscript and contributed to the discussion part of the paper.

CONFLICTS OF INTEREST

The authors declare no conflict of interest.

REFERENCES

1. Shahin, S.M.; Salem, M.A. The challenges of water scarcity and the future of food security in the United Arab Emirates (UAE). *Natural Resour. Conserv.* **2015**, *3*, 1–6.

2. Notarnicola, C.; Caporaso, L.; di Giuseppe, F.; Temimi, M.; Ventura, B.; Zebisch, M. Inferring soil moisture variability in the Mediterrean Sea area using infrared and passive microwave observations. *Can. J. Remote Sens.* **2012**, *38*, 46–59.

3. Sultan, M.; Sturchio, N.; Al Sefry, S.; Milewski, A.; Becker, R.; Nasr, I.; Sagintayev, Z. Geochemical, isotopic, and remote sensing constraints on the origin and evolution of the Rub Al Khali aquifer system, Arabian Peninsula. *J. Hydrol.* **2008**, *356*, 70–83.

4. Giardino, C.; Bresciani, M.; Villa, P.; Martinelli, A. Application of remote sensing in water resource management: The case study of Lake Trasimeno, Italy. *Water Resour. Manag.* **2010**, *24*, 3885–3899.

5. Temimi, M.; Hain, C.R.; Zhan, X.; Rabin, R.; Anderson, M.C.; Notarnicola, C.; Stepinski, J.; Bonhomme, A. On the synergistic use of microwave and infrared satellite observations to monitor soil moisture and flooding. In *Remote Sensing of Energy Fluxes and Soil Moisture Content*; CRC Press: Boca Raton, FL, USA, 2013; pp. 345–362.

6. Cheema, M.J.M.; Bastiaanssen, W.G. Local calibration of remotely sensed rainfall from the TRMM satellite for different periods and spatial scales in the Indus Basin. *Int. J. Remote Sens.* **2012**, *33*, 2603–2627.

7. Fekete, B.M.; Vörösmarty, C.J.; Roads, J.O.; Willmott, C.J. Uncertainties in precipitation and their impacts on runoff estimates. *J. Clim.* **2004**, *17*, 294–304.

8. Liebe, J.; van de Giesen, N.; Andreini, M. Estimation of small reservoir storage capacities in a semi-arid environment: A case study in the Upper East Region of Ghana. *Phys. Chem. Earth Parts A/B/C* **2005**, *30*, 448–454.

9. Temimi, M.; Leconte, R.; Brissette, F.; Chaouch, N. Flood and soil wetness monitoring over the Mackenzie River Basin using AMSR-E 37 GHz brightness temperature. *J. Hydrol.* **2007**, *333*, 317–328.

10. Temimi, M.; Lacava, T.; Lakhankar, T.; Tramutoli, V.; Ghedira, H.; Ata, R.; Khanbilvardi, R. A multi-temporal analysis of AMSR-E data for flood and discharge monitoring during the 2008 flood in Iowa. *Hydrol. Process.* **2011**, *25*, 2623–2634.

11. Sippel, S.J.; Hamilton, S.K.; Melack, J.M.; Choudhury, B.J. Determination of inundation area in the Amazon River floodplain using the SMMR 37 GHz polarization difference. *Remote Sens. Environ.* **1994**, *48*, 70–76.

12. Swenson, S.C. GRACE Monthly Land Water Mass Grids NETCDF RELEASE 5.0. 2012. Available online: https://podaac.jpl.nasa.gov/dataset/TELLUS_LAND_NC_RL05 (accessed on 14 July 2016).

13. Forootan, E.; Rietbroek, R.; Kusche, J.; Sharifi, M.A.; Awange, J.L.; Schmidt, M.; Omondi, P.; Famiglietti, J. Separation of large scale water storage patterns over Iran using GRACE, altimetry and hydrological data. *Remote Sens. Environ.* **2014**, *140*, 580–595.

14. Khalaf, A.; Donoghue, D. Estimating recharge distribution using remote sensing: A case study from the West Bank. *J. Hydrol.* **2012**, *414–415*, 354–363.

15. Gonzalez, A.; Temimi, M.; Khanbilvardi, R. Adjustment to the curve number (NRCS-CN) to account for the vegetation effect on hydrological processes. *Hydrol. Sci. J.* **2015**, *60*, 591–605.

16. Harris, A.; Rahman, S.; Hossain, F.; Yarborough, L.; Bagtzoglou, A.C.; Easson, G. Satellite-based flood modeling using TRMM-based rainfall products. *Sensors* **2007**, *7*, 3416–3427.

17. Smith, L.C.; Pavelsky, T.M. Estimation of river discharge, propagation speed, and hydraulic geometry from space: Lena River, Siberia. *Water Resour. Res.* **2008**, *44*.

18. Khan, S.I.; Yang, H.; Vergara, H.J.; Gourley, J.J.; Brakenridge, G.R.; de Groeve, T.; Flamig, Z.L.; Policelli, F.; Yong, B. Microwave satellite data for hydrologic modeling in Ungauged Basins. *IEEE Geosci. Remote Sens. Lett.* **2012**, *9*, 663–667.

19. Brakenridge, G.R.; Nghiem, S.V.; Anderson, E.; Mic, R. Orbital microwave measurement of river discharge and ice status. *Water Resour. Res.* **2007**, *43*.

20. Smith, L.C.; Isacks, B.L.; Bloom, A.L.; Murray, A.B. Estimation of discharge from three braided rivers using synthetic aperture radar satellite imagery: Potential application to ungaged basins.*Water Resour. Res.* **1996**, *32*, 2021–2034.

21. Sherif, M.; Kacimov, A.; Javadi, A.; Ebraheem, A.A. Modeling groundwater flow and seawater intrusion in the Coastal Aquifer of Wadi Ham, UAE. *Water Resour. Manag.* **2012**, *26*, 751–774.

22. Brook, M.; Al Houqani, C.; Darawsha, T.; Al Alawneh, M.; Achary, S. Groundwater resources; development and management in Abu Dhabi Emirate, UAE. In Proceedings of the 3rd Joint UAE-Japan Symposium, Sustainable GCC Environment and Water Resources, EWR2006, Abu Dhabi, UAE, 28–30 January 2006.

23. Milewski, A.; Sultan, M.; Yan, E.; Becker, R.; Abdeldayem, A.; Soliman, F.; Gelil, K.A. A remote sensing solution for estimating runoff and recharge in arid environments. *J. Hydrol.* **2009**, *373*, 1–14.

24. Al-Dousari, A.; Milewski, A.; Din, S.U.; Ahmed, M. Remote Sensing inputs to SWAT model for groundwater recharge estimates in Kuwait. *Adv. Natural Appl. Sci.* **2010**, *4*, 71–77.

25. Elmahdy, S.I.; Mohamed, M.M. Topographic attributes control groundwater flow and groundwater salinity of Al Ain, UAE: A prediction method using remote sensing and GIS. *J. Environ. Earth Sci.* **2012**, *2*, 1–13.

26. Rizk, Z.S.; Alsharhan, A.S. Water resources in the United Arab Emirates. In *Developments in Water Science*; Abdulrahman, S.A., Warren, W.W., Eds.; Elsevier: Amsterdam, The Netherlands, 2003; Volume 50, pp. 245–264.

27. Böer, B. An introduction to the climate of the United Arab Emirates. *J. Arid Environ.* **1997**, *35*, 3–16.

28. Landerer, F.W.; Swenson, S.C. Accuracy of scaled GRACE terrestrial water storage estimates. *Water Resour. Res.* **2012**, *48*.

29. Kanamitsu, M.; Ebisuzaki, W.; Woollen, J.; Yang, S.-K.; Hnilo, J.; Fiorino, M.; Potter, G.L. NCEP-DOE AMIP-II reanalysis (R-2). *Bull. Am. Meteorol. Soc.* **2002**, *83*, 1631–1643.

30. SCAD. *Energy and Water in Figures*; Statistics Centre: Abu Dhabi, UAE, 2013.

31. Linacre, E.T. Data-sparse estimation of lake evaporation, using a simplified Penman equation. *Agric. For. Meteorol.* **1993**, *64*, 237–256.

32. Allen, R.G.; Pereira, L.S.; Raes, D.; Smith, M. *Crop Evapotranspiration-Guidelines For Computing Crop Water Requirements-FAO Irrigation and Drainage Paper 56* **1998**, *Volume 300*, D05109.

33. Pereira, A.R.; Pruitt, W.O. Adaptation of the Thornthwaite scheme for estimating daily reference evapotranspiration. *Agric. Water Manag.* **2004**, *66*, 251–257.

34. Swenson, S.; Wahr, J. Post-processing removal of correlated errors in GRACE data. *Geophys. Res. Lett.* **2006**, *33*.

35. Page, E. Continuous inspection schemes. *Biometrika* **1954**, *41*, 100–115.

36. Murad, A.; Al Nuaimi, H.; Al Hammadi, M. Comprehensive assessment of water resources in the United Arab Emirates (UAE). *Water Resour. Manag.* **2007**, *21*, 1449–1463.

6
CHAPTER

Wet and Dry Season Effects on Select Soil Nutrient Contents of Upland Farms in North Bank Region of the Gambia

Ebrima Sonko[1*], Daniel N. Tsado[2], Sidat Yaffa[3], Appollonia A. Okhimamhe[1], Julia Eichie[1]

[1]*WASCAL Coordinating Secretariat, Federal University of Technology, Bosso Campus, Minna, Nigeria*

[2]*School of Agriculture and Agricultural Technology, Federal University of Technology, Minna, Nigeria*

[3]*School of Agriculture and Environmental Sciences, University of the Gambia, Brikama Campus, Brikama, The Gambia*

ABSTRACT

The study was conducted in three villages of North Bank Region of the Gambia in 2013 and 2014. We examined wet and dry season effects on select soil nutrient contents of upland farms in North Bank Region of the Gambia. The objective was to evaluate changes in soil nutrient contents in both wet and dry seasons. Soil samples were collected from three RCBD upland fields with three replications at a depth of 0 - 15 cm and analyzed for pH, Soil Organic Carbon (SOC), and soil moisture content. The gravimetric method of moisture estimation was used. The results showed that soil moisture content, soil TN, and soil pH are significantly different ($P < 0.05$) during the two seasons. There was no significant difference in SOC between the two seasons in the study area. The study concluded that soil nutrients were more readily available during the wet season than during the dry season probably because there is more soil moisture available in the wet season that facilitates soil nutrient release. The study concludes that soil moisture has to be available in order for some select soil nutrients to be released for plant uptake.

Keywords: Soil Nutrients, Wet Season, Dry Season, pH

1. INTRODUCTION

Agricultural production systems in The Gambia in general and in the North Bank Region of The Gambia in particular are mainly rainfall-dependent. Most farmers in North Bank Region of The Gambia depend on rainfall for agriculture production. Therefore, when there is drought, North Bank Region becomes highly susceptible to crop failure and other livelihood activities that have a direct bearing on agriculture production. Soils in the study area are generally low in soil moisture as a result of erratic rainfall and also low in fertility due to continuous cropping, without much replenishment of soil nutrients by the farmers. Annual depletion rate of primary nutrients from the soils were estimated at as high as 22 kg of nitrogen (N), 2.5 kg of phosphorus (P) and 15 kg of potassium (K) per hectare over the previous three decades of productive uplands in 37 African nations which include The Gambia (Sanchez et al. 1997) [1] . North Bank Region of The Gambia is the most prone region of the country to drought because of its low vegetative cover. Although there was research conducted in North Bank Region on the impact of drought on crop production that research did not specifically look at the effect of drought on the content of soil nutrients for plant uptake. That research gap is the need for this study to address so that recommendations could be made on adaptation strategies that could be implemented in order to increase crop production. Because of drought challenges in North Bank Region, Yaffa (2013) [2] suggested that more strategies on soil fertility and water conservation methods should be adopted, horticulture in small irrigated gardens be improved, and drought-tolerant rice cultivars be planted as adaptation activities during drought periods.

Soil moisture deficiency has some implications for sustainable agricultural development. Downing and Bakker (2000) [3] confirmed that adverse weather differs from real weather by its (adverse weather) potential to minimize soil moisture availability for plant uptake for growth and development. Keenan and Krannich (1997) [4] reported that drought impact is felt more on societies with erratic rains than other climatic events that may cause soil nutrient deficiency.

Nitrogen, as a major plant nutrient, for example, is an essential component of proteins, coenzymes, and many other metabolic products and is estimated to represent 2% of total plant dry mass (Miller and Cramer, 2004) [5] . Nitrate and ammonium are the primary sources of inorganic N taken up by plant roots but nitrate is often more available to plants for uptake (Hopkins, 1995) [6] . Hopkins (1995) [6] further reported that nitrogen is often delivered to plant roots by mass flow and diffusion, and thus its availability to plants is highly dependent on soil moisture content. Indeed, drought is often associated with nitrogen limitation (Wang et al. 1998) [7] , while excessive water flux can result in nitrogen leaching (Parnes, 1990) [8] .

The availability of nutrients in the soil interface is dependent on the management strategies applied at the field level over decades. On agriculture soils, nutrient depletion is sparse within the land surface because of intensive utilization of fertilizers for export crops. Nutrient-depleted farms in The Gambia as a result of monocropping are still more common than those without nutrient-depleted soils. Nutrient deficiency rates show large differences with soil properties. For example, Pieri (1989) [9] , and Swift et al. (1994) [10] reported that a high rate of leaching occurs in sandy soils than in clay soils because of large pore spaces within soil aggregates in sandy soils.

2. OBJECTIVES

1) To evaluate changes in soil nutrient contents in both wet and dry seasons.

2) Make recommendations on ways to enhance soil moisture availability to release soil nutrients for plant uptake.

3. MATERIALS AND METHODS

3.1. Study Area

The research was conducted in North Bank Region of The Gambia which has a population of about 172,835 people (GBOS, 2003) [11] . It lies between Latitude 13°29'10"N and Longitude 15°5'48"W. The field work took place in only three districts (Central Baddibou District, Jokadu District, and Upper Niumi District). The Region is characterized by erratic rainfall, poor soil structure, and low soil fertility. The climate is of two different seasons: a wet season (that starts in June and ends in October) and a long dry season (that starts in November and ends in May). The average daily temperature is 30°C in the dry season and 27°C in the wet season. Rainfall trends are unpredictable and the annual total is usually less than 800 mm (FAO 2002) [12] . The natural vegetation is mainly savannah woodland and some mangrove trees lining along the banks of River Gambia dissected by an intricate system of water channels. Other areas within the landscape are flood plains and woodlands that are located behind the mangroves.

3.2. Soil Sampling

The locations used for data collection were upland fields in the three Districts. The plots on the fields were used in the study by using a Randomized Complete Block Design (RCBD) with three replications. Soil samples were collected from the replicated plots using a soil probe and screw-type auger. Sukristiyonubowo (2007) [13] method of soil sampling was used to collect soil samples at a depth of 0 - 15 cm and the samples were analysed for pH, Soil Organic Carbon (SOC), Soil Total Nitrogen (TN) and soil moisture content. Soil sampling was done on the 10th and 12th October, 2013 in the wet season and 25th, 26th and 27th January, 2014 in the dry season.

3.3. Soil Analyses

Soil total nitrogen was determined using the micro kjeldahl distillation and titration method described by Bremner, J. M., and C. S. Mulvaney, (1982) [14] . Soil organic carbon (SOC) was determined by a modified Walkley-Black procedure as described by Nelson and Sommers (1982) [15] . The gravimetric method of moisture estimation was used where the soil sample was placed in an oven at 105°C and dried to a constant weight; this took 24-36 hours. Soil pH was determined using the glass electrode method with a soil-to-water ratio of 1:2

3.4. Data Analyses

SPSS software was used in the data analyses. Significant differences between the select soil contents in the two seasons were determined. The significance level was held at $P < 0.05$

4. RESULTS AND DISCUSSION

4.1. Soil Moisture

During the wet season, soil moisture content within the treatment plots fluctuated, which could be attributed to the amount of rainfall and probably the amount of crop biomass at each selected location to conserve moisture in the soil (Figure 1). During the wet season, there was no significant difference ($P < 0.05$) in soil moisture content between the three locations (i.e. 43% at Njaba Kunda, 41% at Jokadu, and 38% at Niumi Lamin, respectively). In the dry season, on the other hand, although soil moisture content was lower than in the wet season in all three locations (Figure 1), still there was no significant difference in between the soil moisture contents. The decrease in soil moisture content could probably be due to high soil moisture evaporation which generally occurs during a drought period. Under an evapo-transpiration rate, utilization of water by crops declines when 50% of the required soil moisture is lost. For example, the water requirement for groundnut production ranges from 500 mm to 700 mm for the production cycle (FAO, 2013) [16] . This shows that the cropping cycle in the North Bank Region could be highly vulnerable to soil moisture loss as a result of drought, which consequently impacts on

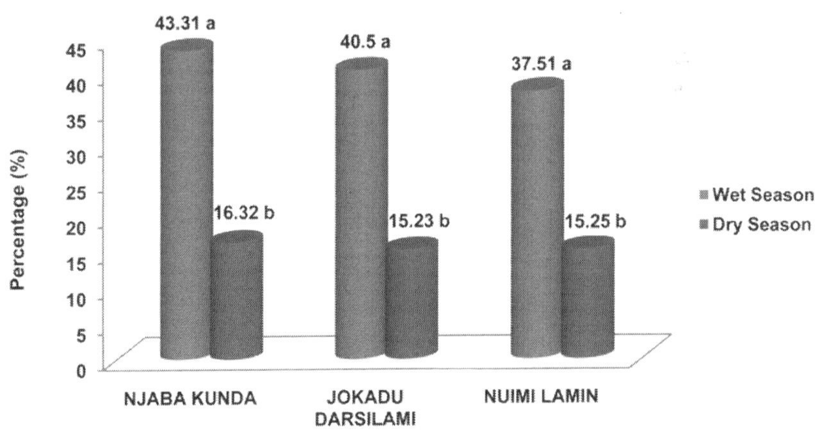

Figure 1. Soil moisture content during wet and dry seasons in three locations in North Bank Region of The Gambia.

crop production and productivity. Water is the primary factor determining soil nutrient mineralization and uptake by crops. Penning de Vries and Djiteye, (1982) [17] reported that Sahel region of Africa highlighted this deficiency in the soil during production seasons. Therefore, crops grown in the North Bank Region could be vulnerable to about 50% water deficiency in the dry season if crop cultivation is also carried out during the dry season.

4.2. Soil pH

The study found no significant difference in soil pH, irrespective of the cropping season, between the three study locations (Figure 2). This is an indication that all the three study locations could probably be of the same soil types that are characterized by low pH. Soil pH has a considerable influence on the availability of nutrients to crops. Low soil pH (i.e. pH between 1 and 4) which is a highly acidic condition tightly holds soil nutrients which are unavailable for plant uptake. Ronen, (2007) [18] reported that most soil nutrients are available for plant uptake in the pH range of 5.5 - 6.5. These low pH values could be the reason for low availability of nutrients for plant uptake.

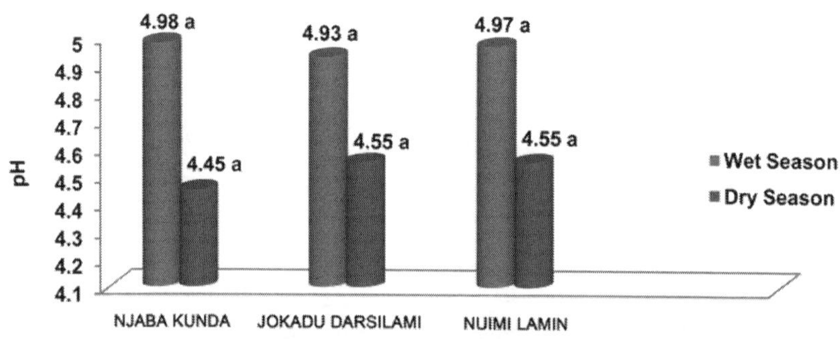

Figure 2. Soil pH content during wet and dry seasons in three locations in North Bank Region of The Gambia.

4.3. Soil Total Nitrogen

Soil total nitrogen recorded for Njaba Kunda, Jokadu, and Niumi Lamin for both wet and dry seasons are shown in Figure 3. In either of the growing seasons, soil total nitrogen was less than 1%. This may be attributed to the low crop residue amount that is left on the farms after crop harvest and or low soil moisture which influences nitrogen mineralization for crop uptake as also reported by Wang et al. (1998) [7] .

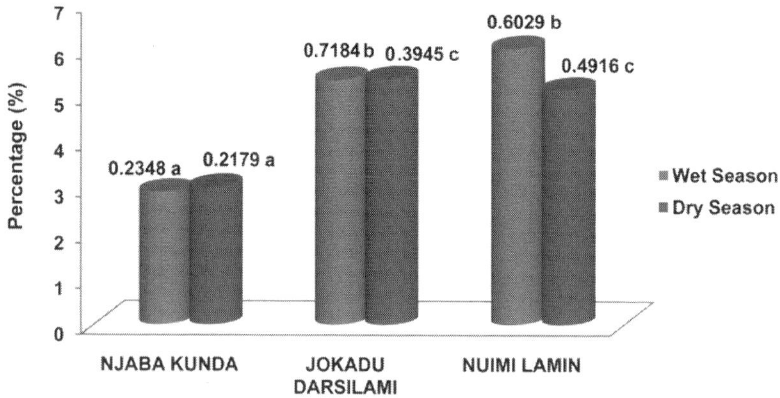

Figure 3. Soil total nitrogen content during wet and dry seasons in three locations in North Bank Region of The Gambia.

Although the soil total nitrogen in either cropping season at either of the three study locations was less than 1%, we found a significant difference in total soil nitrogen between Njaba Kunda and Jokadu Darsilami and Njaba Kunda and Niumi Lamin in either of the seasons (Figure 3). This could be probably due to having more crop residues in Jokadu Darsilami and Niumi Lamin, respectively, than in Njaba Kunda that mineralized and released more soil total nitrogen. In Njaba Kunda, the study found no significant difference in soil total nitrogen between the two seasons (Figure 3). It was difficult to explain this phenomenon because the study was expected to have more soil total nitrogen in the wet season than in the dry season because of more mineralization as a result of more soil moisture availability. In Jokadu Darsilami, the study found a significant difference in soil total nitrogen between the wet and dry seasons. There was about twice more soil total nitrogen in the wet season than in the dry season at Jokadu Darsilami. This was probably that in the wet season, there was more mineralization of nitrogen than in the dry season. In Niumi Lamin, there was about one and a half times more soil total nitrogen found in the wet season than in the dry season (Figure 3). Again, just like in Jokadu Darsilami, probably there was more nitrogen mineralization during the wet season than in the dry season in Niumi Lamin (Figure 3). There was no significant difference in soil total nitrogen, irrespective of the season, between Jokadu Darsilami and Niumi Lamin (Figure 3). This could be that in both Jokadu Darsilami and Niumi Lamin there could have been similar amounts of nitrogen mineralization.

4.4. Soil Organic Carbon

Soil organic carbon content among the three study locations is shown in Figure 4. Irrespective of the study location, there was no significant difference between the two seasons in terms of soil organic carbon content (Figure 4). However, there was a significant difference in soil organic carbon content between Njaba Kunda and Jokadu Darsilami and Najaba Kunda and Niumi Lamin (Figure 4), respectively. This could be probably due to more crop residue found in both Jokadu Darsilami and Niumi Lamin than in Njaba Kunda. Irrespective of the season, there was no significant difference in soil organic carbon content between Jokadu Darsilami and Niumi Lamin (Figure 4). Low soil organic carbon content in the three study locations is supported by findings byWilliams and Balling (1994) [19] which states that African soils, and mostly those of West Africa, have low inherent fertility as a result of low crop residue cover of the soils. The same authors went on further to state that soil fertility depletion (mainly N, P, and C) is the primary factor hindering food security in West Africa.

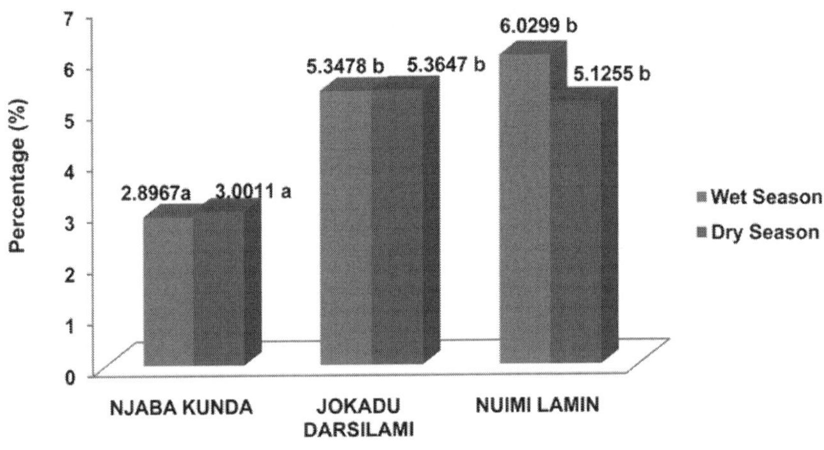

Figure 4. Soil organic carbon content during wet and dry seasons in three locations of North Bank Region of The Gambia.

5. CONCLUSION

Soil moisture content, which ranges between 37.51% and 43.31%, is below optimum soil moisture requirements of 50% for good crop growth and development. The study concludes that the soil pH of both wet and dry seasons are acidic and below the required pH range for crop productivity i.e. 5.5 to 6.5. Meanwhile, soil total nitrogen in the study area was less than

1% in either of the seasons whereas soil organic carbon content ranged between 2.8967% and 6.0299% in both seasons. These were still found to be less than 7% which is ideal for optimum crop production. Therefore, seasonal variation on soil nutrient level has called for sustainable soil fertility management practices that would enhance food security in the North Bank Region of The Gambia, in particular and in The Gambia as a country.

6. RECOMMENDATIONS

1) Do not remove much of the crop residue from the soil surface after crop harvest so that more soil moisture can be retained for plant uptake and nutrient mineralization.

2) Incorporation of crop residues in the soil should be encouraged so that soil carbon and soil nitrogen could be increased.

3) Supplemental crop irrigation should be encouraged in order to avert drought impacts on crop production in the Region.

ACKNOWLEDGEMENTS

This paper is one of the outcomes of masters dissertation submitted to the Federal University of Technology, Minna, Nigeria under the sponsorship of West Africa Science Service Centre on Climate Change and Adapted Land Use (WASCAL), MRP program. The authors are indebted to the following: the Department of Water Resources of The Gambia government for providing rainfall information for the study area, the National Agricultural Research Institute (NARI) of The Gambia government for the technical support rendered during field work, and the farmers of North Bank Region of The Gambia who granted us permission to carryout the study in their fields.

REFERENCES

1. Sanchez, P.A., Shepherd, K.D., Soule, M.J., Place, F.M., Buresh, R.J., Izac, A.-M.N, et al. (1997) Soil Fertility Replenishment in Africa: An Investment in Natural Resource Capital. In: Buresh, R.J., et al., Eds.,

Replenishing Soil fertility in Africa, Soil Science Society of America Spec. Publ. 51. SSSA, Madison, 1-46.

2. Yaffa, S. (2013) Coping Measures Not Enough to Avoid Loss and Damage from Drought in the North Bank Region of The Gambia. International Journal of Global Warming, 5, 467-482.
3. Downing, T.E. and Bakker, K. (2000) Drought Discourse and Vulnerability. Chapter 45. In: Wilhite, D.A., Ed., Drought: A Global Assessment, Natural Hazards and Disasters Series, Routledge Publishers, Abingdon.
4. Keenan, S.P. and Krannich, R.S. (1997) The Social Context of Perceived Drought Vulnerability. Rural Sociology, 62, 69-88.
5. Miller, A.J. and Cramer, M.D. (2004) Root Nitrogen Acquisition and Assimilation. Plant and Soil, 274, 1-36.
6. Hopkins, W.G. (1995) Introduction to Plant Physiology. J. Wiley, New York.
7. Wang, J.R., Hawkins, C.D.B. and Letchford, T. (1998) Photosynthesis, Water, and Nitrogen Use Efficiencies of Four Paper Birch (Betula papyrifera) Populations Grown under Different Soil Moisture and Nutrient Regimes. Forest Ecology and Management, 112, 233-244.
8. Parnes, R. (1990) Fertile Soil: A Grower's Guide to Organic and Inorganic Fertilisers. agAccess, Davis, 190 p.
9. Pieri, C.J.M.G. (1989) Fertility of Soils. A Future for Farming in the West African Savannah. Springer Verlag, Berlin.
10. Swift, M.J., Seward, P.D., Frost, P.G.H., Qureshi, J.N. and Muchena, F.N. (1994) Long Term Experiments in Africa: Developing a Database for Sustainable Land Use under Global Change. In: Leigh, R.A. and Johnson, A.E., Eds., Long-Term Experiments in Agricultural and Ecological Sciences, Centre for Agriculture and Bioscience International, Wallingford, England, 229-251.
11. GBOS (2003) http://www.accessgambia.com/extra/gambia-bureau-of-statistics-gbos.html
12. FAO (2002) Water Control Component of the Special Program for Food Security. TCP/GAM/2801 (D) Project Document.
13. Sukristiyonubowo (2007) Nutrient Balances in Terraced Paddy Fields under Traditional Irrigation in Indonesia. Ph.D. Thesis. Ghent University, Ghent.

14. Bremner, J.M. and Mulvaney, C.S. (1982) Nitrogen-Total. In: Page, A.L., et al., Eds., Methods of Soil Analysis, 2nd Edition, Agronomy Monograph 9. ASA and SSSA, Madisons, 595-624.

15. Nelson, D.W. and Sommer, L.E. (1982) Total Carbon, Organic Carbon and Organic Matter. Methods of Soil Analysis, Part 2. Chemical and Microbiological Properties, 2nd Edition. ASA-SSSA, Madison, 595-579.

16. FAO (2013) Global pressure on Natural Resources Management and Environment Department of FAO in Rome. www.fao.org/nr/water/docs/enduring_farms.pdf

17. Penning de Vries, F.W. and Djiteye, M.A. (1982) La productivité des pasturages Saheliens: Uneétudedes sols, des végétations et de l'exploitation de cette resource naturelle. Agricultural Research Report 918.

18. Ronen, E. (2007) Micro Elements in Agriculture. Practical Hydroponic and Greenhouses, 39-48.

19. Williams, M.A.J. and Balling, R.C. (1994) Interactions of Desertification and Climate. WMO/ UNEP, Geneva, 230 p.

7
CHAPTER

Primitive Geometry Tillage Modeling

Elizabeth Frink, Daniel Flippo

Kansas State University, Manhattan, USA

ABSTRACT

Tillage practices have a significant effect on soil properties. Understanding the specific effects due to geometric and speed parameters of the tillage implement is the key to selecting a tillage practice that is best for a field. This is the first step of many towards optimizing an efficient tillage implement given initial field conditions and desired final conditions. Simple, small-scale tests were performed on idealized implement geometries as a proof-of-concept for future large-scale tests. The results of these tests are presented here and in-depth analysis will be presented in future work.

Keywords: Tillage, Automation, Geometry, Conservation

1. INTRODUCTION

Tillage practices affect soil carbon, water pollution, and farmers' energy and pesticide use, and therefore data on tillage can be valuable to understanding the practice's role in reaching climate and other environmental goals. [1]

The negative effects of tillage have been a driving force behind the popularity of CSV (conservation tillage) and RT (reduced till). Modern NT (no-till) began to emerge with commercial use of synthetic herbicides [2] .

In the 1990s herbicides began to improve, while planting and harvesting equipment designed for NT conditions was developed, leading to the growth of NT farming [3] . The use of herbicides, rather than tillage, leaves the crop residue on the field, which increases the soil's water storage capacity [4] . In addition, conservation tillage (where at least 30% of the soil surface is covered by crop residue) has the potential to reduce fuel consumption, enhance soil microbial activity, increase water infiltration rates, and reduce soil erosion. The impact of CSV and NT is summarized below in Table 1 [5] .

Table 1. Impact of CSV and NT, summarized from [5] .

Impact	Conservation tillage	No-till
Aggregate stability	Organic matter concentrated near surface, encourages microbial growth, increasing aggregate stability.	Crop residue at surface prevents surface crusting, increasing aggregate stability.
Compaction	Improves surface structure due to concentration of residues, decreasing compactibility; Can cause additional compaction in the untilled layer.	Dependent on soil type.
Soil mineralization	Shallower depth with no soil inversion so releases less N for crop uptake than conventional tillage, which leads to less mineralization.	
Emergence and root growth	Decreases soil temperature and increases residue, thus impeding crop emergence; Root growth depends on biological macropore ability to compensate for absence of mechanical macropores.	
Soil water storage and infiltration	Increases C content in soil, which increases water storage capacity and water retention; Infiltration depends on soil type and biological porosity.	
Weeds	Compared to conventional, more weeds in general.	
Disease and pest control	Conventional tillage more effective for control of soil-bourne pathogens, but increased biological activity of conservation and no-till can form disease-suppressive soils.	

A survey conducted at Oklahoma State University [6] revealed perceptions of farmers in Oklahoma, to understand why adoption of NT is lower than the national average. One-third of the farmers who use only one tillage system have tried NT but then switched back to CNV (conventional tillage) within an average of 2.4 years. This switch may be due to a potential for crop yields to decrease in the first two to three years of NT. Additionally, the survey indicated that farmers may not feel informed enough about CSV to be comfortable with switching. The least important perceived benefit appeared to be increased yield, while the most important perceived benefit was reduced fuel costs. From the survey, CNV farmers thought there would be more problems with CSV than those farmers already employing it. The largest perceived problems, from those already using CSV, were lack of state and local research, equipment costs, and lack of knowledge. It is clear that farmers are interested in ways to be more efficient, increase yield, and protect the value of the crop land. However, there is a perceived lack in knowledge and research of tillage practices, and there are advantages and disadvantages to consider with each tillage practice.

There have been numerous experimental studies on the effects of tillage, but not relative to specific geometric parameters of tillage. The existing studies compare conventional tillage to varying types of conservation tillage, but do not look at modifications to the tillage implements. For example, Abdullah [7] found that during germination and flowering, minimum tillage (field cultivator) resulted in higher soil water contents than conventional tillage (moldboard followed by disk). Devita et al. [8] observed similar effects when comparing conventional tillage (moldboard) to no-till, where the soil water content was significantly higher for no-till, leading to more uniform emergence. To varying degrees, several researchers observed the effects of conventional tillage, minimum tillage and no-tillage on bulk density and/or soil porosity [9] - [17] . Penetration resistance relative to tillage practices has also been studied [17] - [20] . Guan et al. [14] observed the effects of plow tillage, rotary tillage and no-till on root weight density. Finally, the effects of various tillage practices on organic content were observed by several researchers [8] [11] [21] . However, none of these studies dealt directly with tillage parameters such a depth, angle of tillage, etc. Additionally, none of these studies have observed all effects on the soil—just selected properties.

This work is the first of many steps towards optimizing tillage implements so that optimal field conditions can be obtained and maintained. As the first step in the process of optimizing tillage implements, a small-scale tillage test system was constructed. The goal of this system was to obtain preliminary data so that the experimental and theoretical modeling processes could be evaluated. Starting with small-scale allowed reduction of the complexity of implement geometry. "Scale models have many advantages, including lower cost of construction, greater flexibility in the range of parameters that can be investigated, and closer control of test conditions." [22] In order to further simplify the evaluation, soil properties were ignored and only a few basic measures of till quality were evaluated. The models developed in this work will be used to optimize implements for desired field measures. The knowledge gained here will facilitate the development of a full-scale system in future work. With a full-scale system, realistic implement geometries and field conditions can be tested, models developed, and implements can be optimized for desired field conditions.

2. METHODS

A variety of implement geometries were tested in a thorough test matrix, using a small-scale setup. The specifics of these are all discussed in the following sections.

2.1. Test Setup

The test bed had internal dimensions of 0.84 m (W) × 1.37 m (L) × 14.0 cm (D), with sand at a depth of 10.2 cm. Extruded aluminum rails constructed the frame of the setup, allowing translations in the x-, y- and z-directions and providing mounting for a camera. Driving motion for the implement was provided by a DC motor actuating a screw drive. Between each test the sand was leveled by sliding a depth-controlled board across the surface of the test bed. The setup is shown inFigure 1.

2.2. Sensors and Controls

Data acquisition and motor control was run with a LabVIEW myRIO (Xilinx Z-7010 FPGA reconfigurable input/output device). A pulse-width modulation pulse from the myRIO controlled a Parallax HB-25 motor controller. Voltage supplied to the motor was measured by the myRIO. A strain gage on the implement was calibrated as a load cell via a P-3500 Strain Indicator module that output continuously to the myRIO. The wiring diagram for all control and acquisition is shown inFigure 2. The LabVIEW VI (virtual instrument) saved all relevant data to a spreadsheet at a rate of 10 Hz.

Figure 1. Small-scale system.

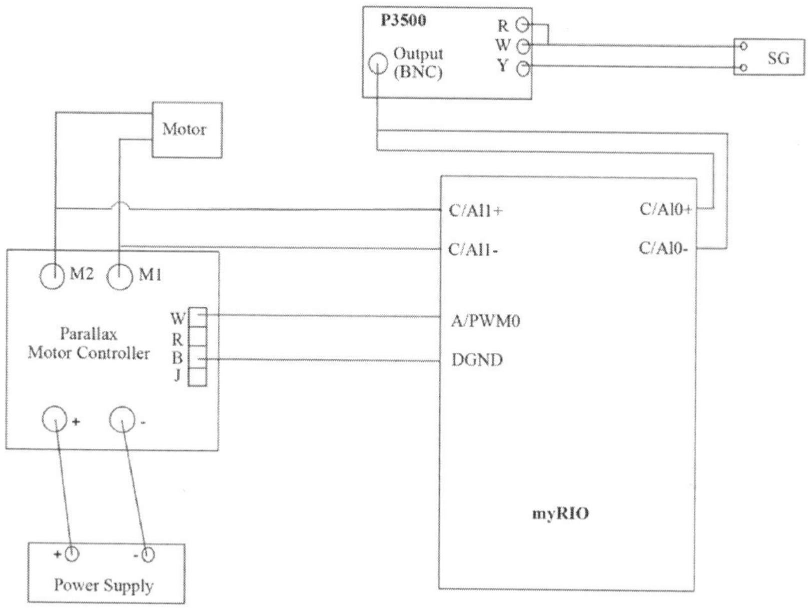

Figure 2. Wiring diagram for data acquisition and motor control.

Additionally, a Logitech HD Webcam C525 was used to capture images of the ridge and furrow after each test. Spotlighting was used to highlight the features in the sand. Post-processing (calibration and measurements) of the images was performed with Image J [23] .

2.3. Test Matrix

Table 2 summarizes the test matrix that was used for each implement. At a depth of 3.0 cm, an input voltage of 6V was not sufficient to drive the plate implement and was thus ignored. Also, a depth of 3.0 cm resulted in too much force for the motor on the sphere and triangle tools and was thus ignored for those two tools.

2.4. Test Implements

Three basic geometries were developed for the test implements, shown in Figure 3. A flat plate, a convex sphere, and an equilateral triangle were used, all with a width of 10.2 cm. These geometries represent the extreme ends of the most simple geometries.

Test Procedure

For a single test, Input Voltage, Depth and Angle were user-specified to identify the test matrix parameters. The screw drive moved the implement through the sand while the VI recorded force continuously, via the calibrated strain gage. After the test was complete, the force was averaged over the range where till of the implement was fully developed (uniform ridge and furrow, constant force). Furrow and ridge measurements were obtained from images taken at the end of each test, using an average of three measurements for each.

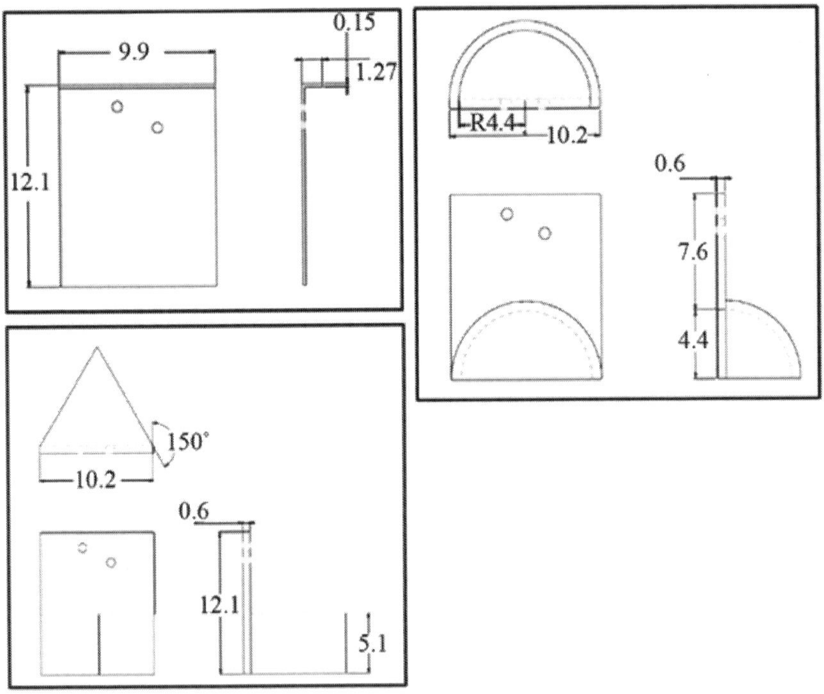

Figure 3. Geometries of the small-scale implements. Units in cm.

Table 2. Small-scale test matrix.

ID	Angle [deg]	Input voltage [V] (speed [cm/s])	Depth [cm]
1	0	6 (0.43)	1.7
2	22.5	7.5 (0.56)	3.0
3	45	9 (0.66)	
4	67.5		

3. RESULTS AND DISCUSSION

Analysis of the results will be discussed in-depth in a later article to facilitate adequate coverage on both topics, and a basic statistical analysis will be presented here. Average and standard deviation values were calculated for each set of three repeated tests, and are also included in Table 3 for plate, Table 4 for sphere, Table 5 for triangle.

Table 6 shows the maximum, minimum and average standard deviation values for the tools. The sphere tool had higher standard deviations than the plate and triangular tool, for all three properties. For all tools, the average standard deviation was higher for the furrow and ridge measurements. This is likely due to the manual measurement method used to obtain these values, compared to the direct measuring of the force values.

Average values for each property are plotted in Figures 4-7 for all of the tools. The average values for the ridge and furrow follow a second order polynomial for the plate but a linear trend for the sphere and triangle. The force values behave exactly opposite. These behaviors are present because of the 3D geometry of the triangle and sphere, where the projected area engaging the sand is not decreasing as significantly as in the case of the flat plate. In the case of the triangle, the projected area actually begins to increase after some rotation.

Equations (1)-(4) give the linear fit for the furrow width (plate tool depth 1.7 cm and 3.0 cm, sphere tool depth 1.7 cm, triangular tool depth 1.7 cm, respectively), where d is the depth of the tool and θ is the angle of the tool. Values for R^2 can be seen in Figures 4-7.

$$furrow_{pl,d=1.7} = -0.0011\theta^2 + 0.22\theta + 4.9043 \quad (1)$$

$$furrow_{pl,d=3.0} = -0.0002\theta^2 - 0.0247\theta + 2.4806 \quad (2)$$

Table 3. Statistics for small-scale tests: 10.2 cm wide, thin, flat rectangular implement.

			Average			Standard deviation		
Input V	Depth [cm]	Angle [deg]	Force [g]	Furrow [cm]	Ridge [cm]	Force [g]	Furrow [cm]	Ridge [cm]
6	1.7	0	34.3	4.93	13.5	1.07	0.04	0.03
7.5	1.7	0	40.2	4.86	13.5	5.00	0.14	0.13
9	1.7	0	44.0	4.88	13.5	2.26	0.23	0.10
6	1.7	22.5	30.5	4.94	12.9	2.01	0.21	0.01
7.5	1.7	22.5	29.0	4.83	12.9	1.24	0.13	0.21
9	1.7	22.5	28.5	4.94	12.7	0.53	0.03	0.01
6	1.7	45	15.7	3.62	10.5	0.23	0.01	0.05
7.5	1.7	45	14.5	3.66	10.5	0.75	0.07	0.04
9	1.7	45	15.1	3.73	10.6	0.73	0.03	0.09
6	1.7	67.5	4.9	1.51	6.6	0.29	0.06	0.10
7.5	1.7	67.5	4.7	1.49	6.7	0.11	0.01	0.00
9	1.7	67.5	4.7	1.54	6.7	0.63	0.06	0.01
7.5	3.0	0	101.0	2.14	15.7	16.85	0.04	0.05
9	3.0	0	101.0	2.65	15.6	17.84	0.04	0.13
7.5	3.0	22.5	72.2	2.07	14.8	0.03	0.24	0.11
9	3.0	22.5	71.8	2.09	14.9	1.64	0.10	0.07
7.5	3.0	45	38.9	0.73	12.7	2.69	0.12	0.07
9	3.0	45	37.0	0.70	12.7	2.06	0.08	0.07
7.5	3.0	67.5	13.3	0.00	8.8	1.19	0.00	0.07
9	3.0	67.5	14.4	0.00	8.8	1.70	0.00	0.08

Table 4. Statistics for small-scale tests: 10.2 cm wide, convex sphere implement.

			Average			Standard deviation		
Input V	Depth [cm]	Angle [deg]	Force [g]	Furrow [cm]	Ridge [cm]	Force [g]	Furrow [cm]	Ridge [cm]
6	1.7	0	361	4.35	15.6	16.9	0.22	0.17
7.5	1.7	0	367	4.45	15.4	4.4	0.07	0.08
9	1.7	0	376	4.56	15.4	18.6	0.03	0.04
6	1.7	22.5	334	4.19	14.9	22.2	0.35	0.15
7.5	1.7	22.5	360	4.46	14.8	24.7	0.03	0.14
9	1.7	22.5	341	4.39	14.8	20.5	0.07	0.06
6	1.7	45	196	3.88	13.5	27.1	0.12	0.11
7.5	1.7	45	195	3.81	13.5	4.1	0.05	0.06
9	1.7	45	202	3.94	13.5	6.2	0.04	0.10
6	1.7	67.5	61	2.71	11.3	4.6	0.07	0.09
7.5	1.7	67.5	66	2.80	11.4	4.0	0.05	0.04
9	1.7	67.5	60	2.79	11.3	1.9	0.10	0.04

Table 5. Statistics for small-scale tests: 10.2 cm wide, equilateral triangle implement.

			Average			Standard deviation		
Input V	Depth [cm]	Angle [deg]	Force [g]	Furrow [cm]	Ridge [cm]	Force [g]	Furrow [cm]	Ridge [cm]
6	1.7	0	0.80	2.47	13.1	0.02	0.21	0.12
7.5	1.7	0	0.84	2.69	13.1	0.02	0.06	0.04
9	1.7	0	0.86	2.26	13.3	0.03	0.70	0.24
6	1.7	22.5	0.74	2.44	12.2	0.01	0.23	0.12
7.5	1.7	22.5	0.82	2.47	12.1	0.01	0.07	0.09
9	1.7	22.5	0.79	2.58	12.1	0.01	0.05	0.07
6	1.7	45	0.44	3.53	12.2	0.02	0.13	0.22
7.5	1.7	45	0.42	3.49	12.1	0.02	0.09	0.13
9	1.7	45	0.42	3.48	12.2	0.01	0.04	0.07
6	1.7	67.5	0.17	4.08	12.7	0.01	0.11	0.12
7.5	1.7	67.5	0.16	4.19	12.7	0.00	0.06	0.02
9	1.7	67.5	0.15	4.26	12.6	0.01	0.07	0.12

Table 6. Standard deviation summary.

		Force [g]	Furrow [cm]	Ridge [cm]
Plate (60 tests)	Max	18	0.25	0.20
	Min	0.0	0.03	0.00
	Avg	4.5	0.08	0.08
Sphere (35 tests)	Max	27	0.36	0.18
	Min	0.0	0.03	0.05
	Avg	14	0.10	0.10
Triangle (36 tests)	Max	14	0.69	0.23
	Min	0.0	0.05	0.03
	Avg	4.5	0.15	0.13

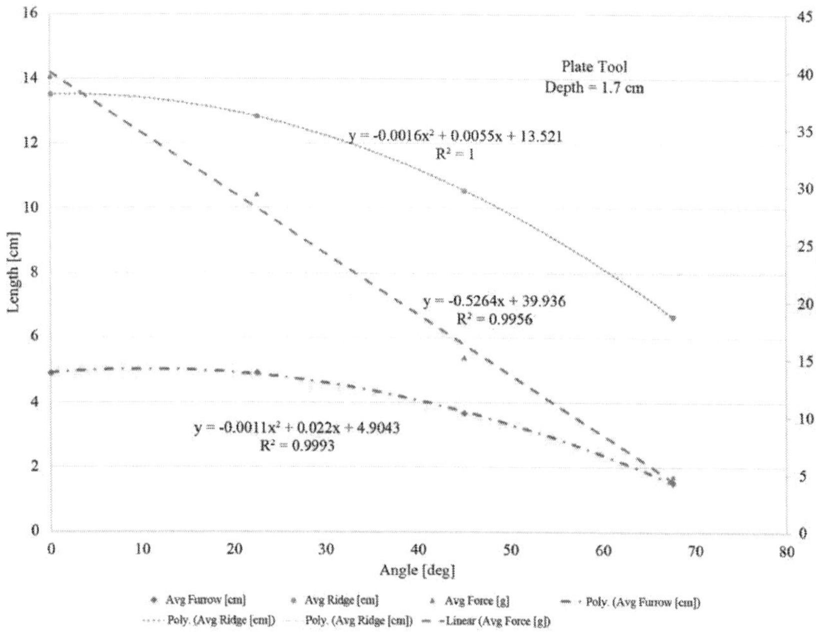

Figure 4. Test results for plate tool at 1.7 cm depth.

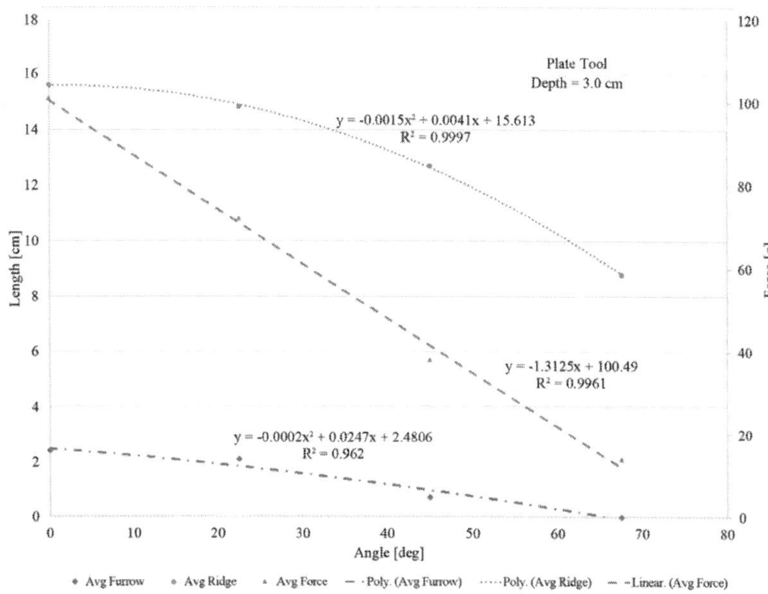

Figure 5. Test results for plate tool at 3.0 cm depth.

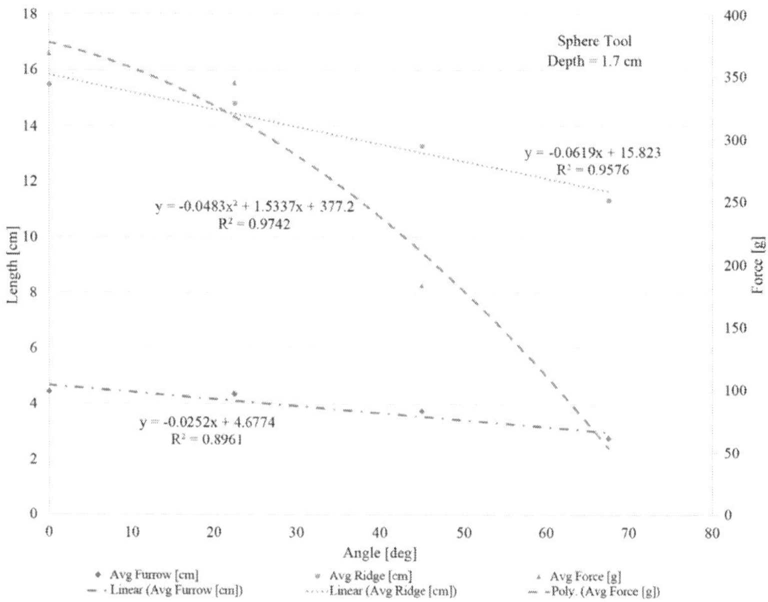

Figure 6. Test results for sphere tool at 1.7 cm depth.

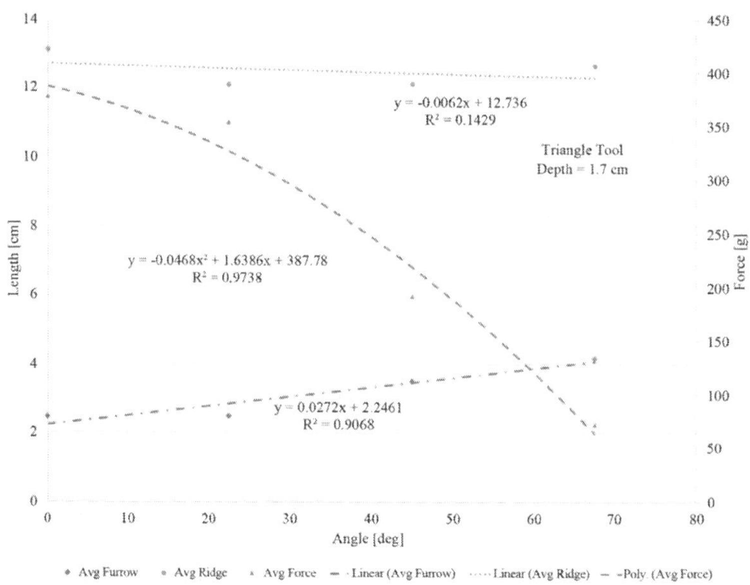

Figure 7. Test results for triangle tool at 1.7 cm depth.

$$furrow_{sph,d=1.7} = -0.0252\theta + 4.6774 \tag{3}$$

$$furrow_{tri,d=1.7} = 0.0272\theta + 2.2461 \tag{4}$$

Equations (5)-(8) give the fits for the ridge width (plate tool depth 1.7 cm and 3.0 cm, sphere tool depth 1.7 cm, triangular tool depth 1.7 cm, respectively). Values for R^2 are in Figures 4-7.

$$ridge_{pl,d=1.7} = -0.0016\theta^2 + 0.0055\theta + 13.521 \tag{5}$$

$$ridge_{pl,d=3.0} = -0.0015\theta^2 + 0.0041\theta + 15.613 \tag{6}$$

$$ridge_{sph,d=1.7} = -0.0619\theta + 15.823 \tag{7}$$

$$ridge_{tri,d=1.7} = -0.0062\theta + 12.736 \tag{8}$$

Equations (9) and (10) give the fits for the plate tool force for depths 1.7 cm and 3.0 cm, respectively. Equation (11) gives the fit for the sphere tool force and Equation (12) gives the fit for the triangular tool force. Values for R^2 are shown in Figures 4-7.

$$F_{pl,d=1.7} = -0.5264\theta + 39.936 \tag{9}$$

$$F_{pl,d=3.0} = -1.3125\theta + 100.49 \tag{10}$$

$$F_{sph,d=1.7} = -0.0483\theta^2 - 1.5337\theta + 377.2 \tag{11}$$

$$F_{tri,d=1.7} = -0.0468\theta^2 - 1.6386\theta + 387.78 \tag{12}$$

4. CONCLUSION

A small-scale test system with a sand bed was successfully implemented. Data for several implement geometries were obtained, following a thorough test matrix for each implement. There results will be used as the foundation for developing a system to optimize a tillage implement. The need clearly exists for improved understanding of tillage. Useful data were obtained and will be analyzed in-depth in future work.

REFERENCES

1. Horowitz, J., Ebel, R. and Ueda, K. (2010) "No-Till" Farming Is a Growing Practice. United States Department of Agriculture, Economic Information Bulletin No. 70.

2. Carr, P.M., Mader, P., Creamer, N.G. and Beeby, J.S. (2011) Editorial: Overview and Comparison of Conservation Tillage Practices and Organic Farming in Europe and North America. Renewable Agriculture and Food Systems, 27, 2-6.

3. Triplett Jr., G.B. and Dick, W.A. (2008) No-Tillage Crop Production: A Revolution in Agriculture! Celebrate the Centennial. A Supplement to Agronomy Journal, S153-S165.

4. Peterson, G.A., Schlegel, A.J., Tanaka, D.L. and Jones, O.R. (1996) Precipitation Use Efficiency as Affected by Cropping and Tillage Systems. Journal of Production Agriculture, 9, 180-186.

5. Peigne, J., Ball, B.C., Roger-Estrade, J. and David, C. (2007) Is Conservation Tillage Suitable for Organiz Farming? A Review. Soil Use and Management, 23, 129-144.

6. Wright, S., Edwards, J., Godsey, C., Vitale, J., Epplin, F. and Taylor, R. (2010) Conservation Tillage in Oklahoma: Percetions and Demographics of Producers.

7. Abdullah, A.S. (2014) Minimum Tillage and Residue Management Increase Soil Water Content, Soil Organic Matter and Canola Seed Yield and Seed Oil Content in the Semiarid Areas of Northern Iraq. Soil and Tillage Research, 144, 150-155.

8. De Vita, P., Di Paolo, E., Fecondo, G., Di Fonzo, N. and Pisante, M. (2007) Notillage and Conventional Tillage Effects on Durum Wheat Yield, Grain Quality and Soil Moisture Content in Southern Italy. Soil and Tillage Research, 92, 69-78.

9. Bhattacharyya, R., Prakash, V., Kundu, S. and Gupta, H.S. (2006) Effect of Tillage and Crop Rotations on Pore Size Distribution and Soil Hydraulic Conductivity in Sandy Clay Loam Soil of the Indian Himalayas. Soil and Tillage Research, 86, 129-140.

10. Chen, H.Q., Hou, R.X., Gong, Y.S., Li, H.W., Fan, M.S. and Kuzyakov, Y. (2009) Effects of 11 Years of Conservation Tillage on Soil Organic Matter Fractions in Wheat Monoculture in Loess Plateau of China. Soil and Tillage Research, 106, 85-94.

11. Dikgwatlhe, S.B., Chen, Z.-D., Lal, R., Zhang, H.-L. and Chen, F. (2014) Changes in Soil Organic Carbon and Nitrogen as Affected by Tillage and Residue Management under Wheat-Maize Cropping System in the North China Plain. Soil and Tillage Research, 144, 110-118.

12. D'Haene, K., Vermang, J., Cornelis, W.M., Leroy, B.L.M., Schiettecatte, W., De Neve, S., Gabriels, D. and Hofman, G. (2008) Reduced Tillage Effects on Physical Properties of Silt Loam Soils Growing Root Crops. Soil and Tillage Research, 99, 279-290.

13. Dal Ferro, N., Sartori, L., Simonetti, G., Berti, A. and Morari, F. (2014) Soil Macro- and Microstructure as Affected by Different Tillage Systems and Their Effects on Maize Root Growth. Soil and Tillage Research, 140, 55-65.

14. Guan, D.H., Zhang, Y.S., Al-Kaisi, M.M., Wang, Q.Y., Zhang, M.C. and Li, Z.H. (2015) Tillage Practices Effect on Root Distribution and Water Use Efficiency of Winter Wheat under Rain-Fed Condition in the North China Plain. Soil and Tillage Research, 146, 286-295.

15. Jabro, J.D., Stevens, W.B., Evans, R.G. and Iversen, W.M. (2009) Tillage Effects on Physical Properties in Two Soil of the Northern Great Plains. Applied Engineering in Agriculture, 25, 377-382.

16. Liu, Y., Gao, M.S., Wu, W., Tanveer, S.K., Wen, X.X. and Liao, Y.C. (2013) The Effects of Conservation Tillage Practices on the Soil Water-Holding Capacity of a Non-Irrigated Apple Orchard in the Loess Plateau, China. Soil and Tillage Research, 130, 7-12.

17. Osunbitan, J.A., Oyedele, D.J. and Adekalu, K.O. (2005) Tillage Effects on Bulk Density, Hydraulic Conductivity and Strength of a Loamy Sand Soil in South-Western Nigeria. Soil and Tillage Research, 82, 57-64.

18. Baumhardt, R.L. and Jones, O.R. (2002) Residue Management and Tillage Effects on Soil-Water Storage and Grain Yield of Dryland Wheat and Sorghum for a Clay Loam in Texas. Soil and Tillage Research, 68, 71-82.

19. Licht, M.A. and Al-Kaisi, M. (2005) Strip-Tillage Effect on Seedbed Soil Temperature and Other Soil Physical Properties. Soil and Tillage Research, 80, 233-249.

20. Munkholm, L.J., Heck, R.J. and Deen, B. (2013) Long-Term Rotation and Tillage Effects on Soil Structure and Crop Yield. Soil and Tillage Research, 127, 85-91.

21. Li, S.W., Jiang, X.J., Wang, X.L. and Wright, A.L. (2015) Tillage Effects on Soil Nitrification and the Dynamic Changes in Nitrifying Microorganisms in a Subtropical Rice-Based Ecosystem: A Long-Term Field Study. Soil and Tillage Research, 150, 132-138.

22. Sullivan, R.J. (1964) Earthmoving in Miniature. National Farm, Construction and Industrial Machinery Meeting, Milwaukee, 14-17 September 1964, SAE Technical Paper 640180.

23. Rasband, W. (2015) ImageJ: Image Processing and Analysis in Java.

8

CHAPTER

Soil Respiration in Drying of an Organic Soil

Zofia Fischer[1], Pavel Blažka[2*]

[1]*Department of Biotechnology and Environment, John Paul II, Catholic University of Lublin, Lublin, Poland*

[2]*Department of Science, University of South Bohemia, České Budějovice, Czech*

ABSTRACT

Drying of soil was linearly related to time, soil volume decreased and ratio of air within the soils increased. Respiration was related with decreasing humidity, storage of CO_2 in soil water results in RQ < 0.5 in the larger soil items at least for a while. Rate of drying decreased in the second part of the process. RQ increased as the CO_2 stored was aerated when its solvent-water evaporated and access of air into the soil increased; eventually RQ = 1.0 in the last days of the experiment. Respiration of the experimental soil stopped when GWC reached 0.15. ΣRQ for the whole process is about 0.7, a bit higher in experiments with less soil suggesting less anoxia.

Keywords: RQ, Anoxia, Storage of CO_2, Changes of Soil Volume

1. INTRODUCTION

In the present contribution our original idea was to study process of drying more in detail. Soils in natural settings are changing; this may be approached not only in field work, but also in lab experiments, where the conditions may be set more accurately and less rely on unpredictable weather etc. and show how they are reflected in the process, particularly in soil respiration. During the work we realized however, that some concepts are generally not quite clear or understood and we propose some clarification, both by pointing to literature not frequently referred in soil science and/or through experimental design.

Respiration was/is considered as a general measure of soil activity (somewhat mystical statement, preferably of soil organisms), recirculation of nutrients from plant residues, while drying of soil is associated with wet and dry periods, seasons, hence more recently with global changes.

Soil respiration was studied many times; the literature on it is quite extensive [1] - [3] . It is based both on field measurements and on laboratory experimentation. The latter is mostly associated with sieving to get more reproducible results, and also with mixing and thus aeration of the soil material. We tried to overcome the sieving effect by slightly compressing the sieved soil by hand [4] . We have seen no other reference on the effect of compression in literature. We want to however first make some introductory comments to the subject respiration. The word respiration is historically (since 18th century) associated primarily with oxygen consumption by men and animals, more recently in biochemistry with oxidation of protons to water in the oxidative phosphorylation pathway as part of the general scheme of energy metabolism in most organisms, prokaryotes and eukaryotes; plants, animals and fungi—except only obligatory anaerobes [5] - [7] . Respiration is a heterotrophic process and also root respiration is not autotrophic e.g. [3] , [8] though it is run by organisms, which largely live autotrophically; all plants are however heterotrophs at night. Soil respiration, meaning respiration of the whole soil community, was introduced later (first half of 20thcentury—Suchtelen [9] was probably among the first). Respiration is anyway a biological process; all biological principles are to be included, not discarded, but it is also a part of soil biology, a composite science as soil physics, biochemistry, geology etc.; all parts have to be observed with the same scrutiny, accuracy and respect to achieve real progress in soil science as a whole.

The sequence/wording “anaerobic respiration” (meaning reduction of e.g. nitrate, nitrite or sulfate replacing dioxygen as proton acceptor) is used quite frequently, though it is not particularly accurate, nitrate and sulfate reduction or proton acceptance would be more to the point, but may be perhaps tolerated.

Respiratory production of carbon dioxide occurs largely in the tricarboxylic acids cycle, is associated through oxidative phosphorylation with oxygen consumption, their ratio in terms of absolute values of moles or volumes is the respiratory quotient (RQ = [production CO_2]/[consumption O_2]).

Introduction of reliable IR non-dispersive spectrophotometers (In fact already 1850 paper based on CO_2) enabling measurements of small concentrations and changes of carbon dioxide concentrations to soil science provoked some shifts in terminology; soil biological activity—measured as production of carbon dioxide—is now quite frequently referred as soil respiration e.g. [2] [8] [10] , most clearly formulated by Gaumont-Guay et al. [11] ("The efflux of carbon dioxide (CO_2) from the soil, referred to hereafter as soil respiration..."). There are however quite a few sources of carbon dioxide in soils, besides processes involving primarily inorganic substrates, e.g. acidification of carbonates by acids [12] , there is also a number of fermentation types occurring when oxygen availability is limited. Both respiration and fermentation are the only processes providing energy to organisms (as macroergic P-bonds [~P], largely ATP). Fermentation in soils contemporary with respiration is apparently quite frequent, particularly when soils become wet and both involve also consumption of some substrate(s); it is therefore surprising, if there is no information on SOM given, particularly in dry environment where very low SOM may be expected e.g. [13] . Carbon dioxide produced in fermentation is associated with much smaller gain in useful energy (about 3 to 4-times) compared to respiration. Respiration supplemented by fermentation is likely the most frequent situation in soils with inadequate oxygen availability; though even low partial tensions of O_2 are mostly adequate for microorganisms. This results in $RQ \geq 1$ and decreasing production of ~P. On the other hand, many products of fermentations may be readily available substrates in respiration, if dioxygen becomes available again. This is also reflected in the interpretation of the RQ. Very often limits for RQ are given e.g. within 0.7 till 1.3. In fact in anaerobes (obligatory or facultative, microbial or eukaryotic) RQ may be much higher (e.g. >10), in strict anoxia it cannot be defined at all, as dividing by 0 (O_2consumption) is a prohibited math operation. Two most important types of fermentation for soil may be lactic and ethanolic (alcoholic). While respiration of lactate and many other products of fermentation (including e.g. acetate) results in $RQ = 1$, like most carbohydrates; for catabolism of ethanol $RQ = 0.57$; in reduction of methane $RQ = 0.50$. Efflux of CO_2 is in fact a mix of respiration, fermentation and abiotic processes leading to absorption or release of CO_2. Field measurements of CO_2 release to the atmosphere are useful particularly for understanding of the role of soils in the global carbon cycle. Experimental conditions (e.g., participation of autotrophs, soil humidity, differences of temperature between the "dome" and the neighboring free area...), are however to be thoroughly specified and the CO_2 efflux is not considered as soil respiration.

In soil literature retention of CO_2 in soil water due its high solubility (distribution coefficient air/water = approx 1 compared to dioxygen 0.03) is mentioned only exceptionally. In soils with pH > 4 dissolved bicarbonate becomes the dominating form of the carbonate system [14] . What we measure by any instrument is the apparent production of CO_2 and RQ modified by diffusion and solubility interference. RQ is a ratio, relationship between the differences of the two components and their ratio are thus non-linear, rather exponential. Jenkins and Adams [15] are thus not correct in saying "RQ is free of empirical assumptions...", Araujo et al. [16] in accepting without comment measured RQ in the range RQ = 0.2 till 1.5 depending on the soil degradation level. On the low side remediation of hydrocarbon degradation of soils results in low RQ [17] , again below the theoretical limit 0.65 for hexadecane; methane is the most reduced hydrocarbon, its oxidation (assuming only one process) results in RQ = 0.5; this should be considered as the low limit of acceptable RQ data. On the other side of possible range synthesis of lipids from carbohydrates are sometimes considered with RQ ≈ 1.3 as possible maximum. It is probably correct for respiration; we must however expect hypoxic or anaerobic conditions particularly in wet soils. Soil organisms are forced to get additional energy to what they get from decreased or even stopped respiration through some fermentation; with prevailing fermentation RQ may be well over 10. In deep anoxia RQ cannot be calculated at all, as CO_2 is produced, but there is no oxygen consumption, division by 0 is a prohibited math operation. Let us just remind summed equation of ethanolic fermentation:

$$C_6H_{12}O_6 = 2\ CH_3 \cdot CH_2OH + 2\ CO_2$$

Being aware of interferences in measurements is particularly important in studies of processes, like various kinds of acclimation to changing temperature and humidity or in the budget of a process.

We want to analyze particular parts of the drying process, not only desiccation, but also changes of soil volume, access of air/oxygen to soil aggregates, storage capacity for CO_2 within the soil etc. and their impact on measured respiration data. This applies both to laboratory and field setups and involves size of soil items in both. Our soil is to some extent exceptional in high SOM content (around 50% of dry weight), many relationships studied in less organic soils need not to be applicable, also some of our conclusions are necessarily not of general use.

Drying of soils is a natural part of regular seasonal events, in observations and/or experiments many papers tend to limit themselves to just momentarily flashes or steady states; we tried however to estimate the budget of the drying process to go through different situations, states of the soil, get rid of many obstacles in getting real metabolic data. Incidentally, we came across low rates, their statistical treatment and time scales.

2. MATERIAL AND METHODS

2.1. The Site and Soil Sampling

The location is on the northern outskirts of the city of Warsaw (Poland). Geographic coordinates are: 52°20'212 N, 020°51'260 E, elevation 81 m. Refer to Blazka and Fischer [18] . The aim of the study is understands of mechanisms of drying through laboratory experimentation and not description of a particular locality. We used the soil for experiments as a laboratory model. The soil is highly organic; nearly 50% LOI, most of the mineral fraction is sand. It comes from a wet site in a dense forest, from the very active A1 zone, depth about 10 cm below surface, it was 5 mm sieved. We got our samples respecting suggestions by Carter and Gregorich [19] .

2.2. Laboratory Analyses

2.2.1. Humidity

(GWC—gravimetric water content, WHC—water holding capacity, dm—dry mass).

Our procedure was based essentially on Grace et al., Priha and Smolander [20] [21] with minor modifications [18] . GWC and other contents are given as ratios, but not %. Dry mass was determined by drying at 110°C to constant weight.

2.2.2. Soil Organic Matter (SOM and Density)

SOM was estimated as loss of weight in ignition to constant weight (LOI) at 500°C on selected samples. Mean densities of SOM (ρ = 1.3) and of the mineral fraction of dry weight (ρ = 2.65), are from Ilsted et al. [4] and many others.

Organic matter = LOI was between 0.45 and 0.50.

ρ(soil) = (LOI * 1.30 + (1 – LOI) * 2.65 = 1.96.

2.2.3. Respiration

(OCD—oxygen consumption, PCO_2—production of CO_2, RQ—respiratory quotient).

We used the volumetric method with the Scholander type respirometer to get both the oxygen consumption and carbon dioxide production. The capacity of its cylindrical flasks is 50 ml (inside radius r = 2 cm, V = 4 cm^2) and we refer data as $\mu l \cdot g^{-1}$(dry mass)$\cdot h^{-1}$. Each experimental series consisted of two parts. In the first, without hydroxide in the flasks, changes in the volumes were recorded mostly for 5 hours, this gave the sum of CO_2 changes and O_2 consumption, in the second part after placing filter paper soaked with NaOH solution into the respirometric chamber, we recorded the oxygen consumption (OCD) for another 5 hours interval, readings each 30 minutes. This involves assumption of slow rates of metabolic changes. The bath temperature was 17.5°C. Sensitivity of the volumetric respirometer is effectively not much worse (<0.1 μl gas·g^{-1}·$hour^{-1}$) than of the IR meters, though rates of providing data are not comparable.

While data of the second part were the OCD, PCO_2 was calculated as first reading (without hydroxide) + second reading (with hydroxide), respecting the signs in readings and calculations. We then plotted the results against time of readings, the slope coefficient of the regression (β_1) was the rate (O_2 consumption-OCD or PCO_2); the respective regression coefficients were all r > 0.95, they were later corrected for pressure and temperature, multiplied by the calibration factor and related per gram dry soil and hour. Weights of respiration vials with soil were taken regularly; tare was subtracted to get rates of drying.

Later in Results and discussion we will consider to what extent the data measured might be biased by diffusion or other transport problems within the soil items, the measured data may then be classified as apparent data (AOCD, $APCO_2$ and ARQ).

2.2.4. Soil (Vs), Air (Va), Pore Volume (Vpo)

Wet soil was gently pressed before start, there were no fissures between soil and wall of respirometric flask, they appeared however later during drying. We measured them by a series of blades (used in fine mechanics to measure narrow crevices, range 0.05 till 2 mm) on four sides of the vial, their mean was subtracted from r = 2 cm to get the radius of the base. The height (he) of the cylinders was estimated using a plate put on the top of the open vial and fixed to center by 4 screws projecting down on the outside of the flask, there were 5 evenly distributed holes on the plate, through which the distance from the top of the plate to soil top was measured by a rod, read by caliper and subtracted from the distance bottom of flask-top of plate. Mean of the 5 readings was height of the soil (he). Final calculation as cylinder volume:

$$\text{Soil volume } (\text{Vs}): \text{Vs} = \pi * r^2 * \text{he} \quad (1)$$

$$\text{Pore volume (porosity) Vpo} = \text{Vs} - ((\text{Vs} - \text{Ws})/\rho) = \text{Vs} - ((\text{Vs} - \text{H}_2\text{O})/\rho) \quad (2)$$

Ws is mass of wet soil.

$$\text{Volume of air Va} = \text{Vpo} - \text{H}_2\text{O} \quad (3)$$

2.3. Experimental

Changes of GWC in experiment are included as background for other parameters of the soil studied. We refer four principal experiments, which were preceded by several preliminary checks or pilot experiments, all done on one and the same kind of soil collected from one spot. All were aimed towards analysis of changes in soil respiration (OCD, PCO_2 and RQ) in drying of the soil. The main difference between the experiments was in the amount of soil in flasks and therefore were coded as LS (less soil) and MS (more soil) respectively. Outline of the experiments is in Table 1. Drying proceeded in MS1 and MS2 in two rates, up to about half of the experiments the rate (a) was somewhat faster, then later (rate b). Drying was terminated when oxygen uptake and CO_2production dropped below the sensitivity of the instrument, except LS1 terminated earlier. There were 12 (LS1, MS1) or later 6 (LS2, MS2) parallel respirometers in each series. The basic data differ by an order of magnitude and change during the experiment; therefore we compared the coefficients of variance-cv's. It turned out that the cv's of the means based on six replicates are smaller than those of 12 replicates (Table 2). As standard deviations (second roots of variance) are involved in calculations of the t-test, we consider the data as comparable (Table 2). Weights of vials for GWC estimates were taken not necessarily on the day of respiration measurement. Therefore GWC data for comparison with respiration were interpolated from the respective regression of GWC to time (all $r > 0.99$).

After each 10 days WHC was estimated in 10 of the vials in the first series (MS1, LS1). In experiment MS2 and LS2 we estimated WHC at the beginning and at termination of the experiment.

Additionally we made a 14-days drying experiment (MP—mud pies) starting at GWC = 1.0 and finished at GWC = 0.13 at the respiration stop. There were two modes of soil drying, each in six respirometers: a) soil (350 g dry weight) dried freely distributed on a plate, aliquots were taken for each respirometric measurements (per 6 * 5 g dry weight), returned to the plate and mixed; b) In the second mode wet soil was pressed by hand into the respirometric vials (per 5 g to each vial), it was left to dry freely between measurements each second day.

We also checked respiration in strict anoxia (SA) comparing respiration of a fresh collected soil with respiration after flushing the respirometers by pure dinitrogen (Air products.com, purity 5.2).

Table 1. Outline of the experiments.

	LS1	LS2	MS1	MS2
dry mass [g] mean; se	3.8; 0.0294	3.26; 0.025	15.22; 0.0751	14.61; 0.147
days of experiment	24	13	52	42
starts of experiment	Aug 25, 14	Feb 23, 15	Nov 11, 14	Feb 23, 15
1st linear section [d]	24	13	32	29
°C air	17.2 - 19.0	18.8 - 19.7	17.0 - 18.8	18.4 - 19.7
β_1 GWC rate a [m]	−1.845	−4.43	−1.632	−1.257
β_1 GWC rate b [m]	x	x	−0.721	−0.412
mud-pies formation [d]	8	6	16	20
humidity air %	50 - 65	40	45 - 62	40 - 45
GWC start	2.4	1.54	2.34	1.58
GWC end	0.644	0.12	0.15	0.16
GWC estimates N	7	5	15	15
GWC replications	12	6	12	6

GWC rates are β_1-coefficients in linear regressions per month, a stands for the first, b for the second linear section. In the LS was one section. LS1 was finished prematurely.

Table 2. Cv's of means of cv's in MS1 and MS2 experiments for OCD, $PCO_2$and GWC to compare variation in data with 12 or 6 replicates.

	OCD	PCO_2	GWC	replicates	determinations
MS1	0.2919	0.3581	0.1275	12	14
MS2	0.1373	0.1401	0.0511	6	13

2.4. Statistics

Was rather basic, it involved averages = means (m), standard deviations (s), coefficients of variation (cv) and standard errors of the mean (se) calculated for all series of data obtained, referred here or not. Two sided t-test was applied for checks of differences of starting and closing data of series and for checks of differences in neighboring data within series. In final evaluation principal (Table 2) rates were related to month: (30days per month).

This came out from the discrepancy in evaluation of rates in changes in GWC, Vs, Vpo related to day which was our working time unit. We were unable to reject the 0-hypothesis (slope-rate of change of GWC is not different from zero—(from horizontal plane line), whereas paired T-test showed significant differences not only between start and end of experiment ($p < 0.001$), but also between most of neighboring data in the series (Table 3). In general the choice of reference units data is frequently intuitional, comparison of two or more statistical treatments might be a more general tool.

3. RESULTS AND DISCUSSION

Results are largely based on the four drying experiments specified in Methods, Experimental, Table 1 and in Figure 1. Additionally we performed two checks to get respiration data of the hard mud pies and the small sandy like aggregates and demonstrate CO_2 production in strict anoxia. Also we assured us that the hard dry mud pies are not just gypsum or carbonate.

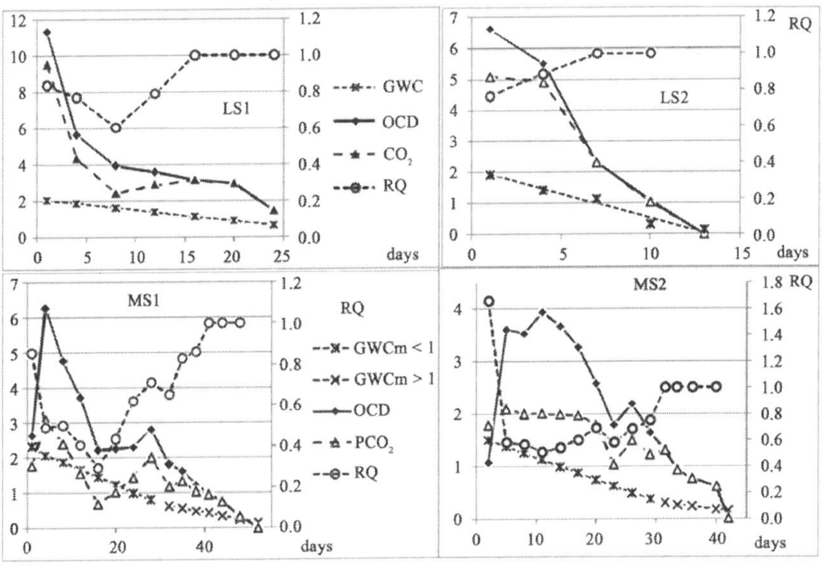

Figure 1. Changes of the gravimetric water content-GWC (as ratios), OCD (oxygen consumption), PCO_2 (production of carbon dioxide)—both as μl·g $[dm]^{-1}·h^{-1}$, and RQ (alternative scales) in the four experiments. Rates of changes in GWC are in Table 1 and Table 3, rates of OCD and PCO_2changes as related to GWC are in Table 5, overall budget of the experiments in Table 6. The LS1 experiment was finished before respiration dropped to zero. All standard errors are within the point icons. Specification of symbols for LS2 and MS2 experiments are in parts for LS1 and MS1 respectively.

Table 3. Rates (slopes) GWC, Vs, OCD per week and month for MS2 and LS2 compared to 2-sided t-test for start-end and neighboring days.

MS2	β_1 (slope)/day	N = 14 all p	slopes/month	p	T-test of pairs start-end	T-test of neighboring days			
						$p > 0.05$	$p < 0.05$	$p < 0.01$	$p < 0.001$
GWC	−0.0361	>0.05	−1.083	<0.005	$p < 0.001$	0		0	14
Vs	−0.4255	>0.05	−12.76	<0.001	$p < 0.001$	6	5	1	2
OCD	−0.0733	>0.05	−2.319	<0.001	$p < 0.001$	3		1	10
LS2		N = 5 for GWC and Vs, N = 4 for OCD							
GWC	−0.1138	>0.05	−3.41	<0.005	$p < 0.001$	0	0	0	5
Vs	−0.3963	>0.05	−11.89	<0.001	$p < 0.001$	0	0	2	3
OCD	−0.5904	>0.05	−17.71	<0.001	$p < 0.001$	0	1	0	3

Significance of slopes related to day and month, compared to 2-sided T-test for data pairs start-end of experiment and neighboring pairs during the experiment.

3.1. Humidity

Relationships of humidity (GWC) are therefore the first to be examined more in detail. Its changes related to time in terms of days[d] proved to be controversial; this resulted in use of month [m] as time reference unit in some selected relationships (Table 3).

GWC decreased during the experiments linearly with time for approximately 2/3 of the course, towards the end in the MS experiments, there was a significant drop in rate (Table 1). The course was a single linear function with respect to time in the shorter LS experiments. A slowing of the function is indicated in LS2, there are, however, too few data to get a significant picture (Figure 1).

3.2. Soil Volume

So far cracks in dry soils were often referred from the field in the literature, but there was/is much less interest in experiments on changes of soil volume in lab. There are more recent works on the impact of repeated drying and rewetting cycles on characteristic of soils resulting so far in non-complete understanding of these processes. We registered shrinking of soil during LS1 and MS1 experiments and estimated it with assumptions of linearity; nevertheless these estimates for MS1 and LS1 are in good agreement with the data from the second series of experiments. All four experiments have shown significant shrinking and changes of pores (porosity). The LS2 experiment is less successful due to high rate of drying and GWC decrease, therefore it was short and only few data were obtained (Table 2 and Table 3).

Soil humidity (GWC) appears to be the governing parameter in shrinking of soil and respiration of soil biota. Shrinking [soil volume Vs, cm^3] and volume of calculated pores, both Vpo and Va, are linearly related with GWC, only Va in LS2 is not significant (Table 4). Regressions of Vs to time [d] were significant.

3.3. Changes of Soil Consistency

Formation of "mud-pies", hard stone—like concretions of dried soil, occurred in all four experiments. In MS1 and MS2 it begun approximately in 1/4 of the drying time, resulted from shrinking of soil, fissures appeared between soil and the walls of respirometric flasks, thus more of the soil surface was opened to the gas phase in

Table 4. Soil volume (Vs), porosity (Vpo) and air filled porosity (Va), their % share in Vs, rates of change (β_1, r), probability of no change (p).

	LS1	LS2	MS1	MS2
Vs; start	9.79 (d8)	8.042	37.7 (d16)	32.735
N; se	9;	9; 0.33	13;	24; 0.509
Vs end	8.17 (d20)	3.035	25.4	15.987
se	x	0.222	x	0.245
Vs-GWC; β_1	1.57	3.74	7.45	11.73
−r	0.944	0.977	0.941	0.990
$p<$	0.025	0.001	0.001	0.001
%Vpo in Vs start	84.2	83.2	87.5	86.38
%Vpo in Vs end	62.3	56.5	50.1	58.32
Vpo-GWC; β_1	2.1775	3.15	10.696	13.252
−r	0.955	0.9947	0.995	0.8767
$p<$	0.001	0.001	0.001	0.001
%Va in Vs start	15.18	16.13	10.3	15.13
%Va in Vs end	36.20	41.79	45.01	68.75
Va-GWC; β_1	−1.13	−0.139	−3.572	−1.468
−r	0.983	0.347	0.964	0.877
p	<0.001	>0.05	<0.001	<0.001

Va did not change significantly in drying, ratio to Vs changed just due to decrease of Vs.

the flasks. The rate of drying was also related to the amount of soil in the flasks. In the MS1 and MS2 series there is a higher soil layer, longer diffusion path and supposed humidity zonation. In the LS series drying and course of the experiments were faster (Table 1); the "mud pies" eventually formed just a narrow plate of dried soil.

All these processes result in increasing access of air into the soil preparation, in a much better ventilation of the dried soils than is in the wet soils and contributed to formation of the "mud pies" during the drying experiments.

It's a common knowledge, that wet soils are soft, pastry; easily adopt form of container under only very slight pressure. When dried they become hard, stone-like, if slightly pressed together before drying; they resist pressures approximately like wall nuts, larger "mud pies" are more resistant than small ones. Sieved soil loosely distributed on a plate acquires in drying a sandy character; its total surface is much greater than that of the "mud pie".

To check the character of the hardiness of the mud pies we chopped off 5.0 g material from the top of the 25 g "mud pie" and took an aliquot of the same weight from the sandy like dried soil. Both were suspended in 50 ml of pure water and extracted for 2 hours and 30 minutes. Both soils looked hydrophobic at the beginning, after approx. 45 minutes they turned more polar. Some air bubbles were attached to dark, presumably light organic soil particles. The extracts were separated from the rests of soils by filtration through Macherey-Nagel filter paper MN 640 m. Both filtrates were yellowish brown, the "mud pie" extract was distinctly darker, but not brown, both very slightly colloidal. Most of the residues remained in the Erlenmeyer's with the heavy sand fraction, much less on the filters. We measured pH, conductivity and redox potential in both extracts: pH = 4.48 and 4.53, conductivity = 54 and 116 $\mu S \cdot cm^{-1}$ in "sand" and "mud pie" respectively, redox was E = 320 mV in both. Addition of 1 ml 0.2 molar solution of $BaCl_2$ to each of the extracts did not result in formation of any precipitate. We conclude, that the top of the "mud pie" probably contained more organic matter than "the sandy" fraction (more intense humic coloration of the extract), there was certainly no carbonate nor gypsum formed in the drying soils, could not been responsible for formation of the "mud pies". The two major components of our soil are the postglacial sand, which however remained unchanged in all the procedures as a separate and non-reactive fraction. The second major component is SOM.

We want suggest replacement of H-bonds (water bridges) in wet soil by some other weak interactions (non- covalent bonds) during drying, possibly by hydrophobic weak bonds. A slight pressure and greater mass of wet soil brings soil aggregates closer together and enables formation of other particle interactions when O-(water-) bridges are abolished in drying.

For considerations of bonding within soil aggregates a change in scale of the aggregates might be useful. Remind that there are 6E23 single molecules per each mole (Avogadro number), thus in 1 μg of SOM 6E14 molecules of organic compounds of mean molecular mass 1000 may be accommodated. Soil science does not seem to be generally ready for thinking in such scales (unfortunate Table 2.1 in Voroney [22] , containing organic molecules [10^{-9} m] larger than viruses, $\Psi > -1500$ kPa not only in Chowdhury [13] etc.).

3.4. Respiration

3.4.1. Relationships to GWC

Relationships of gas exchange (OCD, PCO_2) with time and GWC are more complex, the respective regression coefficients are lower compared to regressions of GWC and time; the overall tendency of respiration (OCD and PCO_2) is a decrease with progressing drying as water is a primary essential-even substrate-of life (Table 5, Figure 1). Data on respiration, both OCD and PCO_2, follow common trends: high values immediately at the begin ning of experiments in both LS, or after three or five days in the MS series. There is a more or less stable respiration in the central part of the course in MS1 and MS2 (Table 2 and Table 4). Basic statistics of respiration data show small variance, se's in all presented and vast majority of data not shown are within the point icons. The data in figures are of similar magnitude with those published by selected authors for forest soils considering problems in recalculation of area related data to mass related. Butler et al. [8] consider "rainfall as main driver of soil CO_2 efflux and its components".

Table 5. Regression coefficients (β_1, r) for respiration to GWC.

r/GWC	LS1	LS2	MS1 all	MS1 after d2	MS2 all	MS2 after d3
OCD β_1	−5.104	−3.612	−1.970	−2.560	−2.000	−2.950
r	0.833	0.956	0.837	0.936	0.712	0.949
PCO_2 β_1	−3.704	−0.877	−0.816	−0.978	−1.149	−1.351
r	0.735	0.953	0.732	0.765	0.842	0.881

For MS experiments coefficients are for all data and also for dates after d2 and d3 due to the delayed Birch effect

3.4.2. CO_2 Storage

Distribution coefficient between gas and water for CO_2 is very close to one at the temperature of the experiments, storage capacity for CO_2 in soil water is therefore equal to GWC * dry mass of soil. It is the volume of water in respirometer (microcosm), about 20 ml in the first days of the MS experiments or 5 ml in the both LS; apparent production in two or three days.

Another estimate of CO_2 stored may result from the difference between AOCD and $APCO_2$ during the flush of respiration when soil is still wet, in MS1 days 3 to12. Assuming real RQ = 1.0 it is 2.81 ml CO_2 stored in 28.8 ml H_2O, the storage capacity is thus used in less than 10%.

Similar data for the MS2 experiment is for days 5 to 12 and show 2.79 ml CO_2 stored in 12.7 ml water, storage capacity used in approximately 22%. Assuming real RQ = 0.7 result in 1.97 ml CO_2stored in MS1 and 1.95 ml CO_2 in MS2 respectively. Details of calculations are not shown.

The storage capacity for CO_2 was falling in experiments with progressing drying. Angert et al. [23] compared measured data of RQ (ARQ) with soil air analysis showing relatively low pCO_2. Their assumption on equilibrium between headspace (soil air) and wet acidic soil might not be fulfilled; there might be physical separation between the sections of the soil gas and the watered section, where CO_2 is primarily produced. Also a careful acidification and mixing of a tightly closed dilute solution of bicarbonate without gas phase need not result in production of gas bubbles. This may apply to a wet soil, where CO_2 is produced by microorganisms. ARQ data of Angert et al. [23] agree with our RQ in the respiration budget, there is also agreement on denial of carbonate chemistry as cause in the background of RQ < 1, we propose oxidation of methane and particularly ethanol produced in fermentation as hypothetic substrates in respiration with RQ < 1. We also seem to agree in accent on processes in studies. Sierra and Renault [14] conclude, that production is negligible compared to dissolved CO_2, this does not apply to acidic soils ($pH < 4$), where bicarbonate is practically missing. Therefore we did not consider bicarbonate storage, as estimates in our soil were always $pH < 4.0$. In acid soils however even a small volume stored may result in serious shifts in RQ.

3.4.3. Anoxia in Soil

To show gas exchange in strictly anoxic soil we placed 21 g fresh collected sieved wet soil (7.3 g dm) into each of four respirometers. Two of the respirometers were flushed with pure dinitrogen 50 $ml \cdot min^{-1}$, for 10 minutes; we registered O_2 consumption and CO_2 production [with and without NaOH trap inside]. The other pair of respirometers was a control to see regular soil respiration. There were 3.83 and 3.89 $\mu l \cdot g^{-1}(dm) \cdot h^{-1}$ respectively gas produced in the N_2 flushed respirometers without the hydroxide trap. There was however no gas production, nor consumption when hydroxide trap was placed in both respirometers; this also shows that the gas produced without the trap is absorbed by hydroxide, it is most probably CO_2 and not CH_4. RQ cannot be determined as shown in Introduction. The regular respiratory control in the second pair of respirometers provided 8.92 and 8.56 µl $O_2 \cdot g^{-1}(dm) \cdot h^{-1}$ respectively and 4.51 µl $CO_2 \cdot g^{-1}(dm) \cdot h^{-1}$ in both respirometers, this is consistent with data in Figure 1.

3.4.4. Respiration and Size of "Mud Pies"

Another pilot experiment aimed to differences of respiration in highly dried soils as "mud pies" or small sandy like aggregates. In the "mud pies" both OCD and PCO_2 are some 20% lower than in the sandy sieved small aggregates. Fast drops of simultaneous decrease of OCD and PCO_2 by approximately 0.6 - 0.8 $\mu l \cdot g^{-1} \cdot h^{-1}$ occurred at 0.63 GWC in "mud pies", at 0.43 in "sand". Drying in "mud pies" and associated decrease of respiration is slower in "mud pies". There is no contradiction to Figure 1.

3.4.5. "Birch Effect"

High respiration data occurring after changed conditions (sieving, watering...) are referred as the Birch effect [24] . In our experiments it occurred in LS1 on d1, in MS1 on d 5 - 14 and also at lower GWC = 1.41 starting value in LS2 and MS2 thus at a somewhat lower moisture throughout the drying experiment (Figure 1). The delayed respiration flush in both MS experiments resulted in distorsions of the relationships of respiration to GWC, this was overcome by skipping the d 1 or d 2 data from the relationships in MS2 and MS1 (Table 6). The delay of flush in both MS experiments was probably caused by pressing the greater amount of soil at start resulting in formation of solid "mud pies", while the small amounts of soil in both LS resulted in a thin layer of soil with minor limitation in gas traffic in the first days. The slightly smaller humidity in MS2 compared to MS1 was of minor importance in this range. We however feel that the impression of extraordinarily of high starting RQ's in MS2 is partly misleading, it comes rather from general characteristics of ratios; OCD and PCO_2 data are of similar magnitude (1.78 and 1.07 respectively), not strikingly different indeed.

High flushes of respiration following change of conditions are debated for last 50 years [24] . So far they were attributed to increased availability of respiration substrates from organisms which died in preceding dries, even if the drop in water content was small [25] or to repeated drying and rewetting (e.g. [26]) or simply to watering of soil (e.g. [13]). Even the detailed study by Unger [27] which includes interaction of soil and the vegetation using highly advanced methodology does not seem to be the last word in the debate. Flush took place in all four experiments in fresh and sieved samples, but not watered, of course sampling was at the very beginning (Figure 1, Table 5, first section of the course). Respiration decreased dramatically in the next readings. Therefore we want to propose a hypothesis not relying primarily on the experimental conditions, but rather on the organisms themselves: Before the collection of soil organisms were adapted to low respiration rates in their natural settings, supplemented by fermentation. Sieving (or in other papers published-water added) allows more air to enter the soil; previously accumulated CO_2 is released to air or absorbed in water. The overall tendency during the experiments is decrease of AOCD and $APCO_2$with decreasing GWC, the data of the first part of experiments are however influenced by CO_2produced and stored in soil water and Va available, at this stage it was not included in the changes of gas contents measured. Some loss of humidity (GWC) in sieving is highly probable. Soil biota may require time to adapt to available dioxygen, presumably by activation/reconstruction of respiratory mechanisms (e.g. enzyme complexes of oxidative phosphorylation pathway) which appear to be of small use in more enzyme copies in wet soils in natural settings with limited oxygen availability.

3.4.6. Summing up OCD and PCO_2 in the Process

Summed OCD and PCO_2 for the whole drying experiment showed in all four experiments RQ around 0.7 (Table 6). ΣRQ is somewhat higher in the faster drying experiments (LS), suggesting less hypoxia. This may support the hypothesis presented in introduction on a mixed status of organisms in wet soils, respiration contemporarily with fermentation and use of the fermentation products (e.g. ethanol) as substrates in respiration when O_2 becomes available.

Studies of some process which take into account all data, characterizing the variable soil condition during a process may provide more information, than short measurements or flashes of soil metabolism in ± constant conditions. Type of instrumentation may be of secondary importance only?modern methods of O_2 estimation are used only exceptionally [27] .

Nevertheless if there is O_2 consumption, anoxia is not in the total mass of soil, possibly in a limited area in center of the soil bloc.

3.4.7. Respiration Ends at Limiting GWC

There is a small, but significant peak of respiration in at the beginning of the decreasing phase at d 26 in MS1; in MS2 at d 26. A similar peak was registered by Thomson et al. [10] on the 5-th day of their 7-days experiment, which did not continue to a respiration stop. We propose to interpret as increased respiration due to increasing access of air to the drying soil, in our experiments shortly overrun but proceeding desiccation of the soil and confirmed by fast drops afterwards.

All four series (including the prematurely finished LS1) end up with RQ = 1 when GWC is approximately = 1 in LS experiments, but GWC ≈ 0.6 in MS (Figure 1).

Table 6. Sums of OCD, COP, and ΣRQ in four experiments [ml].

	LS1	LS2	MS1	MS2
ΣOCD	7.67	15.14	42.53	33.13
ΣCOP	6.31	13.14	23.97	22.41
ΣRQ	0.82	0.87	0.56	0.68

ΣRQ is the ratio of the two sums. Duration of experiments in Table 3.

CO_2 retained in the first section of experiments in water is completely aerated from the soil towards the end when soil is very dry; when respiration seizes at GWC ca 0.15, there is no measurable PCO_2, as is no OCD.

All drying experiments were terminated when GWC dropped to 0.15 and respiration stopped. The experiment LS1 was in fact finished prematurely, nevertheless its final RQ = 1. Also in a preliminary experiment not reported here (summer 2014) respiration (both OCD and PCO_2) seized at GWC = 0.15 and RQ = 1. The respiration limit at GWC = 0.15 need not be a universal value for many soils, Chowdhury et al. [13] reported respiration in Australian soils at recalculated GWC = 0.025.

4. CONCLUSIONS

There is a distinct peak of respiration rate at the beginning of the experiment (Birch effect); it occurred with some delay in the larger soil blocks. The effect was not associated with increased humidity, rather with increased availability of oxygen within the soil; therefore we suggest interpreting as reaction of the enzymatic systems of the soil organisms. The respiration rate decreased linearly with decreasing humidity of the soil; RQ dropped below 0.5 in both larger soil blocks for some time; reason is evidently in storage of CO_2 produced in soil water. Rate of respiration registered is also related to consistency of soil; hard "mud pies" respire some 20% less than the small aggregates of sandy character.

Respiration ceased when GWC dropped down to 0.15. RQ was increasing in the second part of the experiments; a few days before the end it reached 1.00 and remained so till the last moment. We think it comes from CO_2 stored in the earlier parts of the drying process which is aerated when amount of water within the soil decreased and air in soil got higher. ΣRQ for the whole budget of the experiments is about 0.7; it is somewhat higher for the smaller soil items (LS) suggesting less hypoxia and may be consistent with partial use of fermentation products like ethanol (RQ = 0.57).

REFERENCES

1. Davidson, E.A., Savage, K., Verchot, L.V. and Navarro, R. (2002) Minimizing Artifacts and Biases in Chamber-Based Measurements of Soil Respiration. Agricultural and Forest Meteorology, 113, 21-37.
2. Luo, Y. and Zhou, X. (2010) Soil Respiration and the Environment. Academic Press, Burlington
3. Luan, J., Liu, S., Zhu, X., Wang, J. and Liu, K. (2012). Roles of Biotic and Abiotic Variables in Determining Spatial Variation of Soil Respiration in Secondary Oak and Planted Pine Forests. Soil Biology and Biochemistry, 44, 143-150.
4. Ilstedt, U., Nordgren, A. and Malmer, A. (2000) Optimum Soil Water for Soil Respiration before and after Amendment with Glucose in Humid Tropical Acrisols and a Boreal Mor Layer. Soil Biology & Biochemistry, 32, 1591-1599.
5. Campbell, N.A. (1993) Biology. 3rd Edition, Benjamin/Cummings, Francis Group, Redwood City.
6. Alberts, B. and Johnson, A. (2002) The Molecular Biology of the Cell. Garland, Taylor and Francis Group, New York.
7. McGill, W.B. (2007) The Physiology and Biochemistry of Soil Organisms. In: Paul, E.A., Ed., Soil Microbiology Ecology, and Biochemistry, 3rd Edition, Academic Press, Burlington.
8. Butler, A., Meir, P., Saiz, G., Maracahipes, L., Marimon, B.S. and Grace, J. (2012) Annual Variation in Soil Respiration and Its Component Parts in Two Structurally Contrasting Woody Savannas in Central Brazil. Plant Soil, 352, 129-142.
9. van Suchtelen, H. (1910) Uber die Messung der Lebenstatigkeit der aerobiotischen Bakterienim Boden durch die Kohlensaureproduktion. Centrablattfur Bakteriologie, 28, 46-89. Cited from: Lundegardh, H. (1926) Carbon Dioxide Evolution of Soil and Crop Growth. Soil Science, 23, 417-453.
10. Thomson, B.C., Ostle, N.J., McNamara, N.P., Whiteley, A.S. and Griffith, R.I. (2010) Effects of Sieving, Drying and Rewetting upon

Soil Bacterial Community Structure and Respiration Rates. Journal of Microbiological Methods, 83, 69-73.

11. Gaumont-Guay, D.T., Black, T.A., Tim, J., Griffis, T.J., Alan, G., Barr, A.G., Jassal, R.S. and Nesic, Z. (2006) Interpreting the Dependence of Soil Respiration on Soil Temperature and Water Content in a Boreal Aspen Stand. Agricultural and Forest Meteorology, 140, 220-235.

12. Walker, T.S., Harsh, P.B., Grotewold, E. and Vivanco, J.M. (2003) Update on Root Exudation and Rhizosphere Biology. Plant Physiology, 132, 44-51.

13. Chowdhury, N., Burns, R.G. and Marschner, P. (2011) Recovery of Soil Respiration after Drying. Plant Soil, 348, 269-279.

14. Sierra, J. and Renault, P. (1995) Oxygen Consumption by Soil Microorganisms as Affected by Oxygen and Carbon Dioxide Levels. Applied Soil Ecology, 2, 175-184.

15. Jenkins, M.E. and Adams, M.A. (2011) Respiratory Quotients and Q10 of Soil Respiration in Sub-Alpine Australia Reflect Influences of Vegetation Types. Soil Biology & Biochemistry, 43, 1266-1274.

16. Araujo, A.S.F., Cesarz, S., Leite, L.F.C., Borges, C.D., Tsai, S.M. and Eisenhauer, N. (2013) Soil Microbial Properties and Temporal Stability in Degraded and Restored Lands of Northeast Brazil. Soil Biology & Biochemistry, 55, 175-181.

17. Lamy, E., Tran, T.C., Mottelet, S., Pauss, A. and Schoefs, O. (2013) Relationships of Respiratory Quotient to Microbial Biomass and Hydrocarbon Contaminant Degradation during Soil Bioremediation. International Biodeterioration & Biodegradation, 83, 85-91.

18. Blazka, P. and Fischer, Z. (2014) Moisture, Water Holding, Drying and Wetting in Forest Soils. Open Journal of Soil Science, 4, 174-184.

19. Carter, M.R. and Gregorich, E.G., Eds. (2008) Soil Sampling and Methods of Analysis. CRC Press, Taylor & Francis Group, Boca Raton.

20. Grace, C., Hart, M. and Brookes P.C. (2006) Laboratory Manual of the Soil Microbial Biomass Group. Rothamsted Research.

21. Priha, O. and Smolander, A. (2003) Short-Term Uptake of 15NH4 into Soil Microbes and Seedlings of Pine, Spruce and Birch in Potted Soils. Biology and Fertility of Soils, 37, 324-327.

22. Voroney, R.P. (2007) The Soil Habitat. In: Paul, E.A., Ed., Soil Microbiology Ecology and Biochemistry, 3rd Edition, Elsevier, Academic Press, London, 25-49.

23. Angert, A., Yakir, D., Rodeghiero, M., Preisler, Y., Davidson, E.A. and Weiner, T. (2014) Using O2 to Study the Relationships between Soil CO2 Efflux and Soil Respiration. Biogeosciences Discussions, 11, 12039-12068.

24. Birch, H.F. (1964) Mineralisation of Plant Nitrogen Following Alternate Wet and Dry Conditions. Plant Soil, 20, 43-49.

25. Fischer, T. (2009) Substantial Rewetting Phenomena on Soil Respiration Can Be Observed at Low Water Availability. Soil Biology & Biochemistry, 41, 1577-1579.

26. Shi, A. and Marschner, P. (2014) Drying and Rewetting Frequency Influences Cumulative Respiration and Its Distribution over Time in Two Soils with Contrasting Management. Soil Biology & Biochemistry, 72, 172-179.

27. Unger, S., Maguas, C., Pereira, J.S., David, T.S. and Werner, C. (2012) Interpreting Post-Drought Rewetting Effects on Soil and Ecosystem Carbon Dynamics in a Mediterranean Oak Savannah. Agricultural and Forest Meteorology, 154-155, 9-18.

9

CHAPTER

Long-Term Effects of Alternative Residue Management Practices on Soil Water Retention in a Wheat-Soybean, Double-Crop System in Eastern Arkansas

Ryan Norman[1], Kristofor R. Brye[1], Edward E. Gbur[2], Pengyin Chen[1], John Rupe[3]

[1]*Department of Crop, Soil, and Environmental Sciences, University of Arkansas, Fayetteville, USA*

[2]*Agricultural Statistics Laboratory, University of Arkansas, Fayetteville, USA*

[3]*Department of Plant Pathology, University of Arkansas, Fayetteville, USA*

ABSTRACT

Soil water retention is a critical aspect of agricultural management, especially in areas such as the Lower Mississippi River Alluvial Valley that face potential water shortages in the near future. Previous studies have linked changes in soil water retention characteristics to agricultural management practices, especially as they affect the accumulation of soil organic matter (SOM). Therefore, the objective of this study was to determine the relationship between soil water potential and gravimetric soil water content in the top 7.5 cm as affected by nitrogen (N) fertilization/residue level (high and low), residue burning (burning and non-burning), tillage (conventional and no-tillage), and irrigation (irrigated and non-irrigated) after 12 complete cropping cycles in a wheat (Triticum aestivum L.)-soybean [Glycine max (L.) Merr.], double-crop production system in the Delta region of eastern Arkansas using soil wetting curves. The soil investigated was a Calloway silt loam (fine silty, mixed, active, thermic Glossaquic Fraglossudalf). The slope characterizing the relationship between the natural logarithm of the soil water potential and the gravimetric soil water content was only affected ($P < 0.05$) by the N-fertilization/residue-level treatment, while the intercept was unaffected by any field treatment. Averaged across tillage, burning, and irrigation, soil water contents under the high- exceeded those under low-N-fertilization/residue-level treatment at the same water potential, with the greatest differences observed at water contents > 0.12 $g \cdot g^{-1}$. Understanding the ways in which alternative residue management practices affect soil water retention characteristics is an important component of conserving irrigation water resources.

Keywords: Water Retention, Tillage, Residue Burning

1. INTRODUCTION

Agronomic management practices that promote the formation of soil organic matter (SOM) and soil aggregation, such as reduced tillage and diversifying crop rotations, can increase plant available water in the soil [1] and likely have many more positive, long-term effects on soil water characteristics. For example, significant differences have been observed between soil water retention curves in the top 10 cm for native prairie (SOM = 22 $g \cdot kg^{-1}$) and cultivated agricultural soil (SOM = 10.8 $g \cdot kg^{-1}$) in eastern Arkansas [2] . Specifically, the native prairie soil contained a greater soil water content than the cultivated agricultural soil at the same water potential, indicating a possible correlation between increased SOM and water retention. Similarly, decreased soil water retention under conventional tillage (CT) management was reported compared to increased soil water retention and unsaturated hydraulic conductivity under no-tillage (NT) management in a continuous corn (Zea mays L.) study on Mollisols in Iowa [3] . Verkler et al. [4] reported slower soil dry down after wetting under non-burned management compared to burned residue management, as well as slower soil dry down under NT compared to CT when examining soil water content dynamics in a wheat (Triticum aestivum L.)-soybean [Glycine max (L.) Merr.], double-crop system on a silt-loam soil in eastern Arkansas after three years and four complete cropping cycles. Clearly, residue and field management practices influence soil water retention characteristics, which may be related to agricultural management effects on soil aggregation and SOM.

Increases in SOM have been associated with increased infiltration, greater hydraulic conductivity, and increased water retention [5] . Therefore, management practices such as tillage and nitrogen (N) fertilization that may affect the accumulation of SOM may also affect soil water retention characteristics. In a previous study of alternative residue management practice effects on near-surface soil properties in a wheat-soybean, double-crop production system on a silt-loam soil in eastern Arkansas, Amuri et al. [6] reported increasing soil carbon (C) and SOM over time in the top 10 cm across all treatment combinations over the course of six years and seven complete wheat-soybean cropping cycles following conversion to alternative management practices, where trends were likely due to the increase in crop residue returned to the soil as a result of converting from monoculture soybean to a wheat-soybean double-crop system. Smith et al. [7] reported that the abundance of water-stable aggregates was significantly affected by tillage, irrigation, and N-fertilization treatments. Nitrogen fertilization promotes wheat biomass, which may eventually contribute to an increase in SOM and soil aggregation. Therefore, N fertilization, and other management practices that promote SOM and soil aggregation, may affect the relationship between soil water potential and the soil water content, hence soil water retention. For example, Bowman and Halvorson [8] reported significant increases in soil organic C (SOC), and therefore SOM, in the top 5 cm under increased N-fertilization management. Similarly, SOC and SOM increased at a greater rate under a high (134 $kg \cdot N \cdot ha^{-1} \cdot yr^{-1}$) than under low N-rate (<90 $kg \cdot N \cdot ha^{-1} \cdot yr^{-1}$) treatments in a wheat-containing rotation managed consistently for 10 yr near Akron, Colorado [9] .

An understanding of traditional and alternative residue management practices on soil water retention is critical to determining the best management practices in highly productive agricultural regions, especially in areas such as the Delta region of eastern Arkansas that face potential water shortages in the future. Scott et al. [10] used a regression equation based on annual water use rate to determine that 75% of the Alluvial Aquifer, the shallowest aquifer underlying most of the Delta region of eastern Arkansas, will be depleted from large irrigation withdrawls by 2041. Therefore, the objective of this study was to examine the effects of N fertilization/re- sidue level (high and low), residue burning (burning and non-burning), tillage (conventional and no-tillage), and irrigation (irrigated and non-irrigated) on the relationship between soil water potential and gravimetric soil water content in the top 7.5 cm after 12 complete cropping cycles in a wheat-soybean, double-crop production system in eastern Arkansas using soil wetting curves. It was hypothesized that water contents under NT will be greater than that under CT at the same water potential due to effects on aggregation and that the history of irrigation or dryland production will have little effect on soil water retention using the wetting curve approach. It was also hypothesized that the cumulative effects of 12 years of residue burning would render soil water retention in the burned treatment lower than that in the non-burned treatment due to the hydrophobic characteristics of the added ash. In addition, it was hypothesized that increased above and belowground biomass inputs would contribute to greater water retention in the high than in the low N-fertilization/residue-level treatment.

2. MATERIALS AND METHODS

2.1. Site Description

A field study was initiated in Fall 2001 at the Lon Mann Cotton Branch Experiment Station (N34°44'2.26"; W90°45'51.56") [11] in the Lower Mississippi River Alluvial Valley near Marianna, AR. The study site lies within the Southern Mississippi Valley Loess [Major Land Resource Area (MLRA) 134] [12] . This region consists of a series of loess covered hills and alluvial terraces. Despite being prone to erosion, this region has been highly agriculturally productive for decades [12] . The field site is on a Calloway silt loam (fine silty, mixed, active, thermic Glossaquic Fraglossudalf) [13] , which consists of 16% sand, 73% silt, and 11% clay in the top 10 cm [14] . The 30-yr mean air annual temperature of the region is 15.6°C and the 30-yr mean annual precipitation is 128 cm [15] . The 30-yr mean maximum and minimum air temperatures of the region are 32.8°C in July and 2.4°C in January, respectively [15] . This study follows a series of several previous studies at this same study site that have evaluated a variety of short- and long-term effects of alternative management practices effects on plant and soil properties [6] [7] [16] .

2.2. Experimental Design

Between Fall 2001 and Spring 2005, the study consisted of a three-factor, split-strip-plot, randomized complete block experimental design with six replications of each of eight treatment combinations [11] . The three experimental factors were i) N fertilization/residue level (i.e., a high N fertilization/residue level, achieved with a split application of N fertilizer, and a low N/fertilization/residue level, achieved with minimal to no N additions); ii) residue burning (i.e., burning or non-burning); and iii) tillage (i.e., CT or NT) [11] . However, an irrigation factor was introduced in 2005 by dividing the site into an irrigated half and a non-irrigated half [4] . Since 2005, the experimental area has consisted of 48, 3- × 6-m plots with six replications for every N-fertilization/residue-lev- el-burning-tillage treatment combination and three replications for every N-fertilization/residue-level-burning- tillage-irrigation treatment combination (Figure 1) [6] .

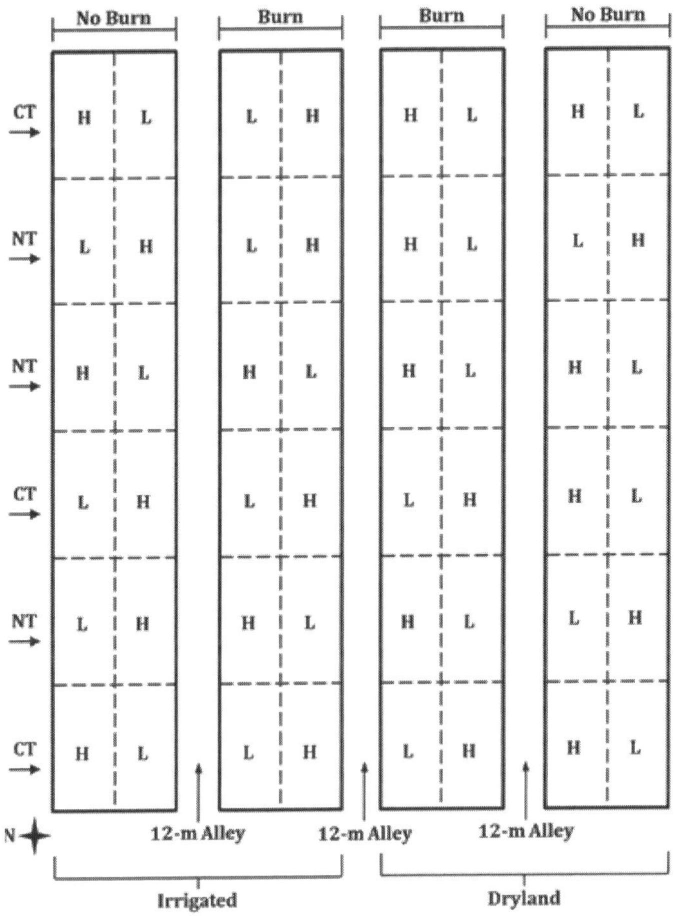

Figure 1. Experimental layout at the Lon Mann Cotton Branch Experiment Station in eastern Arkansas depicting 48, 3- × 6-m plots under residue-level [high (H) and low (L)], burn, tillage [conventional tillage (CT) and no-tillage (NT)], and irrigation treatments.

2.3. Field Management

Prior to the initiation of the study, the site was managed as a continuous, mono-cropped soybean system using CT [11] . Due to the consistent field management prior to beginning this long-term study in 2001, near-surface soil properties throughout entire study area were assumed uniform and any subsequent observed differences in measured soil and/or plant properties were assumed be to the result of imposed field treatments rather than inherent differences among plots [14] .

To prepare for this study, initial field preparations in Fall 2001 involved disking twice followed by broadcast applications of N, phosphorous, potassium, and pelletized limestone at rates of 20, 22.5, 56, and 1120 $kg{\cdot}ha^{-1}$, respectively, prior to wheat planting. Wheat was drill seeded with a 19-cm row spacing each fall thereafter. All plots were manually broadcast fertilized in early March 2002 through 2004 with urea (46% N) at the rate of 101 $kg{\cdot}N{\cdot}ha^{-1}$. High-N-fertilization/residue-level plots ($n = 24$) were manually broadcast fertilized in late March at approximately the late-jointing stage with an additional 101 $kg{\cdot}N{\cdot}ha^{-1}$ to produce different levels of wheat residue. No N-fertilizer was applied in Spring 2005 due to a failure to establish wheat stands caused by prolonged wet soil conditions in Fall 2004. Since 2006, initial applications of 56 $kg{\cdot}N{\cdot}ha^{-1}$ as urea were broadcast on high- N-fertilization/residue-level plots in approximately late February, followed by a split application of an additional 56 $kg{\cdot}N{\cdot}ha^{-1}$ at the late-jointing stage in approximately late March. Since 2006, the low-N-fertilization/residue- level plots have received no N fertilization in order to achieve the desired residue-level difference.

Wheat was harvested using a plot combine in approximately early June each year. Wheat residue left behind the plot combine was uniformly spread by hand over each plot immediately following wheat harvest. Any remaining wheat stubble was mowed with a rotary mower to a height of ~3 cm from the soil surface in order to achieve a uniform residue-covered surface for soybean planting. Following mowing, the burning treatment was imposed on half of the plots by propane flaming. The residue-burning treatment was not able to be imposed in 2005, 2007, and 2012 due to the absence of a wheat stand in Spring 2005, prolonged wet soil conditions in Spring 2007, and overly weedy conditions in 2012. Imposition of the burning treatment was followed by impos ing the tillage treatment each year. The CT plots were disked at least twice with a tandem disk to a depth of ~10 cm followed by seedbed smoothing with at least three passes of a soil conditioner. The CT practices used in this study are representative of widely used pre-soybean-planting tillage operations throughout the region.

A glyphosate-resistant soybean cultivar, maturity group 5.3 or 5.4, was drill-seeded with 19-cm row spacing at a rate of approximately 47 kg seed

ha^{-1} in early to mid-June each year. Potassium fertilizer was applied according to recommended rates [17] when the previous year's soil test indicated potassium was needed. In 2002 through 2004, all plots were furrow-irrigated as needed, three to four times each soybean-growing season. A levee was created in 2005 to exclude furrow-irrigation water from the non-irrigated (i.e., dryland) treatment, which received only natural rainfall. Weeds and insects were managed consistently throughout the entire study area as necessary based on University of Arkansas Cooperative Extension Service's recommendations, which generally consisted of herbicide and insecticide applications during both the wheat- and soybean-growing seasons [17] . Soybean were harvested with a plot combine from late October to early November each year. Each year within two weeks of soybean harvest, the subsequent wheat crop was sown into the soybean residue, which was left in place without any manipulations.

2.4. Plant Sample Collection and Processing

To verify achievement of the N-fertilization/residue-level treatment, immediately following wheat harvest and rotary mowing and before residue burning each year, aboveground residue was assessed. A residue sample was collected from within a 0.25-m^2 metal frame, oven-dried for 3 to 7 days at 55°C, and weighed to obtain an estimate of aboveground residue mass into which soybean were subsequently planted.

2.5. Soil Sample Collection and Processing

Consistent with prior annual soil sampling conducted at this field site [6] [7] [14] [16] , at wheat harvest in May 2014, a single soil sample was collected from the top 10 cm in each plot using a 4.8-cm-diameter stainless steel core chamber beveled to the outside to minimize compaction. Soil samples were oven-dried for 48 hr at 70°C, weighed for bulk density determinations, ground to pass through a 2-mm mesh screen, and analyzed for selected soil chemical properties. Soil pH was determined potentiometrically in a 1:2 (w/v) soil-to-water suspension. Soil OM was determined by weight-loss-on-ignition after 2 hr at 360°C. Total soil C and N were determined by high-temperature combustion with an Elementar VarioMAX Total C and N Analyzer (Elementar Americas Inc., Mt. Laurel, NJ). All measured soil C was assumed to be organic given the lack of effervescence upon treatment with dilute hydrochloric acid [14] . The soil C: N ratio was calculated from measured C and N concentrations. Total C and N and SOM contents ($kg \cdot m^{-2}$) were calculated from measured concentrations ($g \cdot kg^{-1}$) and measured bulk densities in the top 10 cm from the May 2014 soil sampling.

To assess field treatment effects on the relationship between soil water potential and soil water content, 12, ~2-cm-diameter soil samples were collected from the top 7.5 cm of each plot at wheat harvest in May 2014 and combined into one composite soil sample per plot. Each sample was manually homogenized and air-dried for approximately 5 d, ground, and sieved to pass through a 2-mm mesh screen. Subsamples were weighed, oven-dried at 70°C for 48 hr, and reweighed to obtain the initial moisture content of the air-dried sample. Following the procedures of Brye [2] , seven, 5 ± 0.01-g subsamples of air-dried soil from each of the 48 plots were added to small mixing cups. Drops of distilled water (i.e., 2, 4, 6, 10, 12, 15, and 20 drops) were added to each of the seven mixing cups with an eyedropper and manually mixed with a spatula to achieve a range of gravimetric soil water contents. The moist soil in each mixing cup was transferred to small plastic instrument cups, 4 cm in diameter by 1 cm tall, and lightly packed to a uniform bulk density of ~0.7 $g{\cdot}cm^{-3}$. The plastic instrument cups were capped and allowed to equilibrate overnight to room temperature (i.e., ~20°C). The next morning after over-night temperature equilibration, the water potential was subsequently measured with a WP4 Dewpoint PotentiaMeter (Decagon Devices, Inc., Pullman, WA), which was calibrated using a standard potassium chloride solution. After the water potential was recorded, each instrument cup was weighed, oven-dried at 70°C for 48 hr, then reweighed for gravimetric water content determinations. Measured water potentials were natural-logarithm transformed to facilitate statistical analyses.

In addition, to determine actual residue management effects on near-surface bulk density, a third set of soil cores were collected between 8 and 10 weeks after soybean planting in 2014, similar to previous annual bulk densities assessments [6] [7] [14] [16] . A single 4.8-cm-diameter soil core was extracted from the top 10 cm in each plot using a stainless steel core chamber and slidehammer. Samples were oven-dried at 70°C for 48 hr, and weighed for an additional bulk density measurement.

2.6. Statistical Analyses

An analysis of variance (ANOVA) was conducted using SAS (version 9.3, SAS Institute, Inc., Cary, NC) to evaluate the N-fertilization/residue-level treatment effect on aboveground residue mass for annual measurements conducted between 2007 and 2014. In addition, an ANOVA was also conducted using SAS, based on the strip- split-plot experimental layout of the field treatments (Figure 1), to evaluate the effects of N fertilization/residue level, burning, tillage, irrigation, and their interactions on soil pH, SOM, total N, and C contents, C:N ratio, and bulk density in the top 10 cm measured in 2014, which represents the cumulative effects of 12 consecutive years of consistent management. Due to practical limitations of the study area, the addition of the irrigation treatment since 2005 was superimposed on the burning treatment (Figure 1). Therefore, irrigation and burning treatments could not be simultaneously analyzed within this experimental design. Therefore, two separate ANOVAs were conducted, each excluding one of the confounding factors (i.e., burning and irrigation). When appropriate, treatment means were also separated by least significant difference (LSD) at the 0.05 level.

To evaluate field treatment effects on soil water retention, an analysis of covariance (ANCOVA) was conducted using SAS to examine the long-term effects of N fertilization/residue level, burning, tillage, irrigation, and their interactions on the linear relationship between the natural logarithm (log) of the measured soil water potential and the gravimetric water content from the soil wetting-curve data. For the purposes of simultaneously evaluating all field treatments and their interactions, the experimental design was assumed to be completely random with three replications of each of 16 treatment combinations. The full ANCOVA model was reduced using a hierarchal principle to remove non-significant terms, and non-significant terms were only included the final model when they participated in higher-order, complex treatment interactions. When appropriate, treatment means for slopes and intercepts from the log-transformed relationships were separated by LSD at the 0.05 level.

3. RESULTS AND DISCUSSION

3.1. Aboveground Residue Levels

Following the 13 wheat crops produced in this study between 2002 and 2014, the N-fertilization scheme used annually produced numerically greater aboveground residue amounts, into which the subsequent soybean crop was planted, in the high than in the low N-fertilization/residue-level treatment in 12 of the 13 years, with 2002 being the exception [6] [18] . In addition, significantly ($P < 0.05$) greater aboveground residue amounts were produced in the high than in the low N-fertilization/residue-level treatment in 10 of the 13 years, with 2002, 2004 [6] , and 2010 being the exceptions [18] . Therefore, it is clear that the intended residue-level differences were achieved in more than the majority of the 13 years to justify the N-fertilization/residue-level treatment.

3.2. 2014 Soil Properties

After 13 complete wheat-soybean cropping cycles (i.e., 2001 to 2014) and 12 years of consistent management, soil C and N contents, soil C:N ratio, and soil pH in the top 10 cm were affected ($P < 0.05$) by field treatments. When irrigation was excluded from the model, soil C content ($P = 0.038$) and soil C: N ratio ($P = 0.033$) differed between burn treatments in 2014. Averaged across tillage, N fertilization/residue level, and irrigation, soil C content averaged 1.22 and 1.42 kg·m^{-2}, while the soil C:N ratio averaged 9.2 and 10.2 under burning and non- burning, respectively. However, when burning was excluded from the model, soil N content ($P = 0.032$) and the C:N ratio ($P = 0.021$) differed between the N-fertilizer/residue-level treatments (Table 1). Averaged across tillage, burning, and irrigation, soil N content averaged 0.14 and 0.13 kg·m^{-2} under the high and low N-fertiliza- tion/residue-level treatments, respectively. Similarly, averaged across tillage, burning, and irrigation, the soil C: N ratio averaged 9.4 and 10.0 under the high and low N-fertilization/residue-level treatments, respectively.

In 2014, soil pH in the top 10 cm differed ($P = 0.021$) between irrigation treatments within N-fertilizer/resi- due-level treatments (Table 1) when burning was excluded from the model. Averaged across tillage and burning, soil pH was greater under irrigation regardless of N-fertilization/residue level, where soil pH averaged 7.26 under the high and 7.28 under the low N fertilization/residue level, than that under the dryland treatment, where soil pH averaged 6.67 under the low, which was greater than that under the high N fertilization/residue level (i.e., pH averaged 6.48). However, all soil pH values, regardless of management, exceeded the minimum soil pH threshold of 6.0, below which soybean yield reductions can be expected on silt-loam soils in eastern Arkansas [19] . Therefore, the differences in soil pH among irrigation and N-fertilization/residue-level treatment combinations were agronomically non-significant with regards to soybean production on silt-loam soils in eastern Arkansas.

In contrast to other initial soil properties, after 13 complete wheat-soybean cropping cycles (i.e., 2001 to 2014) and 12 years of consistent management, bulk density and SOM contents were unaffected ($P > 0.05$) by any of the field treatments in 2014 regardless of whether burning (Table 1) or irrigation (data not shown) were excluded in the statistical model. Therefore, bulk density averaged 1.21 g·cm^{-3} [standard error (SE) = 0.01] and SOM content averaged 2.9 kg·m^{-2} (SE = 0.06) across all field treatments in 2014.

Table 1. Analysis of variance summary of the effects of N-fertilization/residue level (residue level), tillage, irrigation, and their interactions on soil bulk density, pH, soil organic matter (SOM), total carbon (TC), total nitrogen (TN), and the soil C: N ratio in the top 10 cm from spring 2014 after 12 complete cropping cycles in a wheat-soybean, double-crop production system in eastern Arkansas.

Source of Variation	Bulk Density	pH	SOM	TC	TN	C:N
				P		
Residue level	ns[†]	ns	ns	ns	0.032	0.021
Tillage	ns	ns	ns	ns	ns	ns
Irrigation	ns	0.023	ns	ns	ns	ns
Residue level × tillage	ns	ns	ns	ns	ns	ns
Residue level × irrigation	ns	0.021	ns	ns	ns	ns
Tillage × irrigation	ns	ns	ns	ns	ns	ns
Residue level × tillage × irrigation	ns	ns	ns	ns	ns	ns

[†]Not significant (ns, $P > 0.05$).

3.3. Soil Water Retention

As was expected, the relationship between soil water potential and gravimetric water content across all data and treatment combinations followed a curvilinear pattern, where the water potential increased exponentially as gravimetric soil water content increased (Figure 2). Though soil properties, and water retention, were assumed to be uniform at the beginning of the study site in 2001, after 13 complete wheat-soybean cropping cycles (i.e., 2001 to 2014) and 12 years of consistent management, the slope characterizing the linear relationship between the natural logarithm of water potential and gravimetric water content in the top 7.5 cm was affected ($P = 0.007$) by only the N-fertilization/residue-level treatment (Table 2; Figure 3), and was unaffected ($P > 0.05$) by tillage, burning, irrigation or any of their interactions (Table 3). The intercept characterizing the linear relationship between the natural logarithm of water potential and gravimetric water content was unaffected ($P > 0.05$) by any field treatment (Table 2; Figure 3). Averaged across tillage, burning, and irrigation, the slope for the low was greater (i.e., more negative) than that for the high N-fertilization/residue-level treatment (Table 3). This result indicated that, on average, there was a greater change in water content per unit change in water potential associated with the soil from the low compared to the high N-fertilization/residue-level treatment. The greatest differences between high and low N-fertilization/residue treatments were observed at water contents > ~0.12 $g \cdot g^{-1}$ (Figure 3). Conversely, as soil water potential decreased, gravimetric soil water contents became increasingly similar under both N-fertilization/residue-levels treatments.

The results of this study on a silt-loam, loessial soil from the Delta region of eastern Arkansas were similar to the soil moisture characteristic curve results reported by Brye [2] using a similar wetting-curve approach. Brye [2] demonstrated that soil water contents in both native prairie and cultivated agricultural silt-loam soils in east- central Arkansas became increasingly similar as soil water potential approached permanent wilting point (i.e., −1.5 MPa), regardless of field treatments imposed. Verkler et al. [4] also reported numerically greater maximum soil water contents at the 7.5 cm depth under the high compared with the low N-fertilization/residue-level treatment, although the differences were statistically non-significant. Management practices that increase the amount of crop residue returned to the soil, such as with greater above- and belowground biomass achieved with differential N fertilization, have been shown to increase infiltration, bulk density, and water storage capacity [20] .

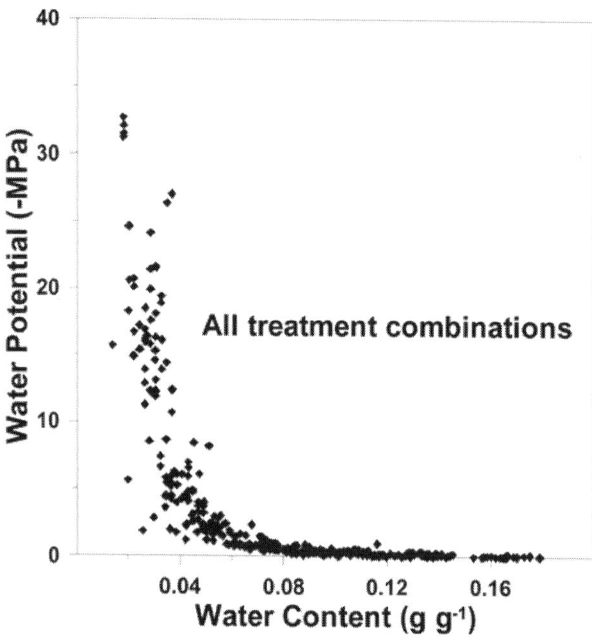

Figure 2. Raw data from all treatment combinations depicting the relationship between soil water potential and gravimetric soil water content from soil wetting curves for the top 7.5 cm in a wheat-soybean, double-crop system in eastern Arkansas after 13 years (i.e., 2001 to 2014) of consistent residue management.

Table 2. Analysis of covariance summary of the effects of N-fertilization/residue level (residue level), burning (burn), tillage, irrigation, and their interactions on the linear relationship between the natural logarithm of soil water potential and gravimetric soil water content using a soil wetting curve approach from the top 7.5 cm after 12 complete cropping cycles in a wheat-soybean, double-crop production system in eastern Arkansas.

Source of variation†	P
Intercept term	
Residue level	ns††
Burn	ns
Tillage	ns
Irrigation	ns
Residue level × burn	ns
Residue level × tillage	ns
Residue level × irrigation	ns
Burn × tillage	ns
Burn × irrigation	ns
Tillage × irrigation	ns
Residue level × burn × tillage	ns
Residue level × burn × irrigation	ns
Residue level × tillage × irrigation	ns
Burn × tillage × irrigation	ns
Residue level × burn × tillage × irrigation	ns
Liner term	
Water content	<0.001
Residue level × water content	0.007
Burn × water content	ns
Tillage × water content	ns
Irrigation × water content	ns
Residue level × burn × water content	ns
Residue level × tillage × water content	ns
Residue level × irrigation × water content	ns
Burn × tillage × water content	ns
Burn × irrigation × water content	ns
Tillage × irrigation × water content	ns
Residue level × burn × tillage × water content	ns
Residue level × burn × irrigation × water content	ns
Residue level × tillage × irrigation × water content	ns
Burn × tillage × irrigation × water content	ns
Residue level × burn × tillage × irrigation × water content	ns

†Non-significant interactions ($P > 0.05$) were removed in the final model, except when non-significant terms participated in higher-order, complex treatment combinations. ††Not significant (ns, $P > 0.05$).

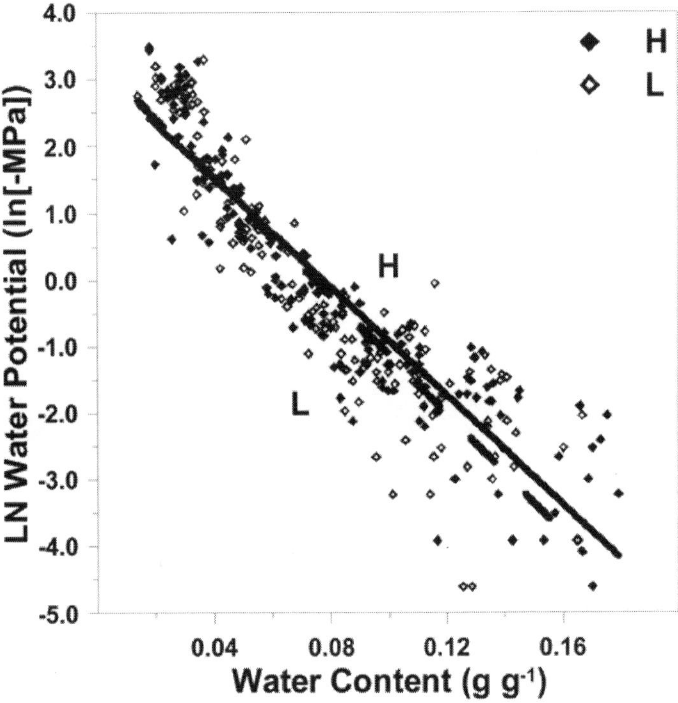

Figure 3. Influence of residue level [high (H, —) and low (L, -----)] on the relationship between the natural logarithm (LN) of the water potential and the gravimetric soil water content from soil wetting curves for the top 7.5 cm in a wheat-soybean, double-crop system in eastern Arkansas after 13 years (i.e., 2001 to 2014) of consistent residue management.

Table 3. Summary of estimated regression parameters for the N-fertilization/residue-level [High (H) and Low (L)] treatment for the linear relationship between the natural logarithm of soil water potential (−MPa) and gravimetric soil water content ($g \cdot g^{-1}$) in the top 7.5 cm after 12 complete cropping cycles in a wheat-soybean, double-crop production system in eastern Arkansas. Coefficient estimates with the same lower case letter do not differ (P > 0.05).

Regression Term	Treatment	Coefficient Estimate	P†
Intercept	High	3.108 a	<0.001
	Low	3.357 a	0.002
Linear	High	−39.712 a	<0.001
	Low	−45.207 b	0.001

†$P < 0.05$ indicates coefficient estimate was significantly different from 0.

Though N fertilization/residue level did not significantly affect SOM or C contents in the top 10 cm in 2014 after 13 years of consistent residue management, one possible explanation for the significant effect of N fertilization/residue level on the relationship between the natural logarithm of water potential and the gravimetric soil water content was that the high N-fertilization/residue-level treatment promoted increased soil structure development and SOM more than the low N-fertilization/residue-level treatment. While it was concluded in a previous study analyzing soil properties in the same plots used in the current study that N fertilization/residue level alone had no obvious, observable effects on the trend in SOM content in the top 10 cm over time between 2007 and 2014, N fertilization/residue level did affect ($P < 0.05$) the trend in SOM content over time as part of complex treatment combinations [18] .

It is also possible that the N-fertilization/residue-level treatment may have impacted SOM content and soil aggregates in the top 7.5 cm differently than in the top 10 cm due to the greater accumulation of both above- and below-ground plant biomass concentrated near the soil surface. A previous study analyzing soil aggregation in the top 10 cm in the same plots used in the current study reported that the concentration of water-stable aggregates was 11% greater in the top 5 cm than in the 5 to 10 cm depth interval after 7 years of consistent management [7] suggesting that SOM and soil aggregates may be more concentrated in the 7.5-cm depth samples used for the current study than in the 10-cm depth samples used for previous studies [6] [18] . Therefore, it is possible that the N-fertilization/residue-level treatment affected the <2-mm-sized soil aggregates, which may have occluded SOM, in the top 7.5 cm, without clearly and obviously affecting SOM contained in the aggregate size classes larger than 2 mm in the top 10 cm. Such an increase in occluded SOM in smaller aggregates might account for an increase in soil water content [5] [21] . For example, Brye [2] reported greater soil water contents in the top 10 cm of a native prairie soil than soil water contents of cultivated agricultural soils at the same water potential and ascribed at least a partial explanation to the greater SOM content in the prairie soil (SOM = 22 $g \cdot kg^{-1}$) compared wo the cultivated agricultural soil (SOM = 10.8 $g \cdot kg^{-1}$) examined.

Somewhat surprisingly, neither burning nor tillage had an observable effect ($P > 0.05$; Table 2) on the relationship between the natural logarithm of water potential and the gravimetric soil water content after 13 years of consistent residue management and 12 complete cropping cycles. These results were in contrast with the original hypothesis that tillage would strongly affect the relationship between soil water potential and soil water content, such that soil water content under NT would exceed that under CT at the same water potential. Similarly, these results were in contrast to the original hypothesis that residue burning would render soil retention in the burned treatment lower than that in the non-burned treatment due to cumulative effects of hydrophobic ash additions. It is possible that any potential effects of burning and/or tillage on water retention characteristics were masked as a result of the sample preparation procedure (i.e., grinding and sieving to ≤2 mm), which rendered the soil structure and natural aggregation and porosities highly disrupted from their in-situ, undisturbed state. As hypothesized, the cumulative effects of irrigation or dryland soybean production had no effect on resulting water retention characteristics as determined using soil wetting curves.

In contrast to the results of this study, other soil water retention studies have reported significant correlations between cultivation and near-surface soil water retention characteristics [2] [3] [5] , likely due to the increased hydraulic conductivity and infiltration rates associated with relatively undisturbed NT compared to highly disturbed CT soils. Verkler et al. [4] reported that residue burning significantly affected maximum soil water contents during irrigation events, and that the mean maximum soil water content was 3% (v/v) greater under residue burning compared with non-burning. However, the water content measurements conducted by Verkler et al. [4] were made in-situ (i.e., in undisturbed soil in the field), and are therefore fundamentally different from the oven- dried, ground, sieved, and rewetted soil samples used in the current study.

4. CONCLUSION

Following conversion to alternative residue and water management practices and after 13 consecutive years (i.e., 2001 to 2014) of management, N fertilization/residue level, but not burning, tillage, or irrigation treatments, significantly affected the linear relationship between the natural logarithm of water potential and the gravimetric soil water content as determined by soil wetting curves. The wetting-curve procedure provided a simple, replicatable, and useful approach to evaluate long-term effects of various field treatment on soil water retention. It can be inferred from the results of this study that differences in N-fertilization/residue-level management can have a cumulative effect on soil water retention characteristics, possibly as a result of the increased soil aggregation and occluded SOM associated with increased crop residue inputs, both above- and belowground, under a high compared to a low residue-level management scheme. Consideration of soil water retention characteristics is vital to planning sustainable use of irrigation water, especially in areas such as the Delta region of eastern Arkansas that will face potential water shortages in the near future.

ACKNOWLEDGEMENTS

Field assistance provided by Matt Gregory at the Cotton Branch Experiment Station is gratefully acknowledged. This work was funded by the Arkansas Soybean Research and Promotion board.

REFERENCES

1. Nielsen, D.C., Vigil, M.F., Anderson, R.L., Bowman, R.A., Benjamin, J.G. and Halvorson, A.D. (2002) Cropping System Influence on Planting Water Content and Yield of Winter Wheat. Agronomy Journal, 94, 962-967.

2. Brye, K.R. (2003) Long-Term Effects of Cultivation on Particle Size and Water-Retention Characteristics Determined using Wetting Curves. Soil Science, 168, 459-468.

3. Hill, R.L., Horton, R. and Cruse, R.M. (1985) Tillage Effects on Soil Water Retention and Pore Size Distribution of Two Mollisols. Soil Science Society of America Journal, 49, 1264-1270.

4. Verkler, T.L., Brye, K.R., Gbur, E.E., Popp, J.H. and Amuri, N. (2008) Residue Management and Water Delivery Effects on Season-Long Surface Soil Water Dynamics in Soybean. Soil Science, 173, 444-455.

5. Azooz, R.H. and Arshad, M.A. (1996) Soil Infiltration and Hydraulic Conductivity under Long-Term No-Tillage and Conventional Tillage Systems. Canadian Journal of Soil Science, 76, 143-152.

6. Amuri, N., Brye, K.R., Gbur, E.E., Popp, J. and Chen, P. (2008) Soil Property and Soybean Yield Trends in Response to Alternative Wheat Residue Management Practices in a Wheat-Soybean, Double-Crop Production System in Eastern Arkansas. Journal of Integrative Biosciences, 6, 64-86.

7. Smith, S.F., Brye, K.R., Gbur, E.E., Chen, P. and Korth, K. (2014) Residue and Water Management Effects on Aggregate Stability and Aggregate-Associated Carbon and Nitrogen in a Wheat-Soybean, Double-Crop System. Soil Science Society of America Journal, 78, 1378-1391.

8. Bowman, R.A. and Halvorson, A.D. (1998) Soil Chemical Changes after Nine Years of Differential N Fertilization in a No-Till Dryland Wheat-Corn-Fallow Rotation. Soil Science, 163, 241-247.

9. Halvorson, A.D., Reule, C.A. and Follett, R.F. (1999) Nitrogen Fertilization Effects on Soil Carbon and Nitrogen in a Dryland Cropping System. Soil Science Society of America Journal, 63, 912-917.

10. Scott, H.D., Ferguson, J.A., Hanson, L., Fugitt, T. and Smith, E. (1998) Agricultural Water Management in the Mississippi Delta Region of Arkansas. Arkansas Agricultural Experiment Station, Division of Agriculture, University of Arkansas, Fayetteville.

11. Cordell, M.L., Brye, K.R., Longer, D.E. and Gbur, E.E. (2006) Residue Management Practice Effects on Soybean Establishment and Growth in a Young Wheat-Soybean Double-Cropping System. Journal of Sustainable Agriculture, 29, 97-120.

12. Brye, K.R., Mersiovsky, E., Hernandez, L. and Ward, L. (2013) Soils of Arkansas. Arkansas Agricultural Experiment Station, University of Arkansas System Division of Agriculture, Fayetteville, AR. 136 p.

13. Natural Resources Conservation Service (NRCS), Soil Survey Staff (2012) Web Soil Survey: Soil Data Mart.

14. Brye, K.R., Cordell, M.L., Longer, D.E. and Gbur, E.E. (2006) Residue Management Practice Effects on Soil Surface Properties in a Young Wheat-Soybean Double-Crop System. Journal of Sustainable Agriculture, 29, 121-150.

15. National Oceanic and Atmospheric Administration (NOAA) (2002) Climatography of the United States No. 81, Monthly Station Normals of Temperature, Precipitation, and Heating and Cooling Degree Days 1971-2000: Arkansas. United States Department of Commerce, National Climatic Data Center, Asheville.

16. Verkler, T.L., Brye, K.R., Popp, J.H., Gbur, E.E., Chen, P. and Amuri, N. (2009) Soil Properties, Soybean Response, and Economic Return as Affected by Residue and Water Management Practices. Journal of Sustainable Agriculture, 33, 716-744.

17. 17. University of Arkansas Cooperative Extension Service (UACES) (2000) Arkansas Soybean Handbook. Miscellaneous Publication 197. Cooperative Extension Service, University of Arkansas, Little Rock.

18. Norman, C.R. (2015) Long-Term Effects of Alternative Residue Management Practices on Near-Surface Soil Properties and

Soybean Production in a Wheat-Soybean, Double-Crop System in Eastern Arkansas. M.S. Thesis, University of Arkansas, Fayetteville.

19. Slaton, N., Roberts, T. and Ross, J. (2013) Fertilization and Liming Practices. In: Extension Soybean Commodity Committee, Ed., Arkansas Soybean Production Handbook, University of Arkansas Research and Extension, Little Rock.

20. Shaver, T.M., Peterson, G.A., Ahuja, L.R., Westfall, D.G., Sherrod, L.A. and Dunn, G. (2002) Surface Soil Physical Properties after Twelve Years of Dryland No-Till Management. Soil Science Society of America Journal, 66, 1296-1303.

21. Dao, T.H. (1993) Tillage and Winter Wheat Residue Management Effects on Water Infiltration and Storage. Soil Science Society of America Journal, 57, 158-159.

10
CHAPTER

Planting Cotton in a Crop Residue in a Semiarid Climate: Water Balance and Lint Yield

Robert J. Lascano[1], Dan R. Krieg[2], Jeffrey T. Baker[3], Timothy S. Goebel[1], Dennis C. Gitz III[1]

[1]*Wind Erosion and Water Conservation, USDA-ARS*, Lubbock, TX, USA*

[2]*Department of Plant and Soil Science, Texas Tech University, Lubbock, TX, USA*

[3]*Wind Erosion and Water Conservation, USDA-ARS, Big Spring, TX, USA*

ABSTRACT

Cotton (Gossypium hirsutum L.) is planted on more land area than any other crop on the Texas High Plains. Much of this area is considered highly erodible and requires a conservation compliance program to participate in government farm programs. Because this region is semiarid and because irrigation water is increasingly limited, water conservation and efficient use of water are necessary to maximize cotton lint yields. One popular conservation compliance practice used is to plant cotton into a chemically terminated small grain crop, i.e., residue that provides wind protection to the cotton seedlings. Our hypothesis was that in a semiarid region the use of a small grain cover crop under irrigated conditions would use more water than it conserves compared to conventional tilled cotton, thus reducing cotton lint yields. To test the hypothesis separate field studies over two growing seasons and on two soil textures, a loamy fine sand and a clay loam, were conducted. The main treatments were tillage systems (conventional and conservation using terminated wheat residue). The two split plot treatments were water supply based on replacement of calculated grass reference evapotranspiration (ET_o). Tillage did not affect the amount of water used by the cotton crop at either location (< 7% difference, P >

0.05) except for an 80% ET_o irrigation treatment at a single location where the bare soil treatment used 10% more water than the residue treatments for both years. The residue treatment decreased ($P < 0.05$) cotton lint yields at both locations by 12% except for the 50% ET_o single irrigation treatment in which the residue treatment yielded 14% more lint than the bare soil treatment. The use of terminated wheat residue had no impact on soil water storage during any part of the year. During a 5-month period associated with wheat growth, the wheat evapotranspiration was 20 to 40 mm more water ($P < 0.05$) than that lost through soil water evaporation from the conventional treatments. The use of terminated wheat residue did not benefit the water balance of the cotton crop, and was associated with decreased cotton lint yields. The results were consistent with our working hypothesis, and disproved the idea that planting cotton into wheat stubble cover increases water use efficiency.

Keywords: Soil Water, Evapotranspiration, Cropping Systems, Semi-Arid, Wheat Residue

1. INTRODUCTION

In 2013 and 2014, about 1.2 million ha were planted with upland cotton on the Texas South Plains [1] . The annual mean precipitation in this region is 467 ± 159 mm and about 50% of this rainfall occurs during the 4-month (May to August) growing season. About half of the planted area is irrigated with a water supply ranging from 2.5 to 7.5 $mm \cdot d^{-1}$ and the potential evaporative demand is on average 9.8 ± 0.8 $mm \cdot d^{-1}$ during the four-month growing season [2] . The source of the irrigation water is the Ogallala aquifer, where withdrawals exceed natural recharge, resulting in a declining water table [3] . Much of the water applied to the cotton is lost to bare soil water evaporation (E_{soil}) in conventional tillage systems [4] [5] . In this semiarid region, where cotton is planted in a wide row spacing (1-m) the crop seldom reaches a leaf area index of 3 $m^2 \cdot m^{-2}$ that is necessary to achieve canopy closure [6] [7] . Furthermore this exposed and bare soil is conducive to increase the amount of E_{soil} [8] .

Given the decline of irrigation water from the Ogallala aquifer [3] [9] with a decreasing pumping capacity, particularly during periods of low rainfall which are frequent in this region, led to develop agronomic practices to maximize cotton's water use efficiency (WUE), i.e., production of lint yield per unit of water taken up by the plant [10] . An example of this conservation practice is to plant cotton into a winter wheat (Triticum aestivum L.) crop that is chemically terminated in the spring before the cotton crop is planted [11] . Residues may protect cotton seedlings by providing shelter to wind damage and sand abrasion [12] . Also, the residue may reduce E_{soil}; however, in some cases it may use more water than it conserves [5] [8] via wicking [13] . This is the process whereby the wheat residues may contribute to additional water loss via capillary action, i.e., wicking effect [14] . In this situation, the wheat residue may increase the ability to transport water from the seedbed to the atmosphere via evaporation and thus resulting in a reduction of soil water and negatively affecting cotton emergence when planted in the wheat residue.

In the cotton cropping systems of the Texas High Plains, the effect of crop residues on water conservation and lint yield, are a function of the water supply, from irrigation and rainfall, soil type, and tillage practices [15] [16] . Planting cotton into a wheat residue on a sandy loam soil only increased cotton lint yield when the crop was irrigated at the grass reference evapotranspiration demand (ET_o). However, this practice is not suited to the availability of irrigation water from the Ogallala aquifer. In field trialsand over a five-year period, in a loamy fine sandy soil the residue did not impact lint yield under irrigated conditions, but significantly and adversely affected dryland lint yields. In a clay loam soil, the lint yield of cotton planted in a wheat residue increased on average by 13% under irrigation and reduced dryland cotton lint yield by 8% [17] [18] . From these results, it can be concluded that the effects of wheat residues on cotton lint yield depend primarily on the water supply. The texture effect is related to the amount of water that can be stored in the soil, particularly from the time the wheat is terminated to when the cotton is planted. For example, in the surface 0.20 m of the soil profile, a clay soil can store about twice as much water as a sandy soil and for dryland conditions the additional stored water would be available during cotton planting and favor seed germination and emergence of the cotton plant.

In first analysis for the Texas High Plains, the amount of water required to establish the wheat residue is often ignored in the water balance of dryland and limited irrigated cotton cropping systems [4] [5] [8] . To establish a wheat residue that would not impact the germination and emergence of the subsequent cotton crop planted in the residue, and of sufficient growth to be beneficial to cotton seedlings requires a net gain of water. In the Texas High Plains, winter wheat is planted in months (November to February) where the average monthly rainfall is 16 ± 2 mm [2] . Clearly, in a dryland system the establishment of the residue comes with the risk of using stored water that otherwise would be available at the time needed when the cotton crop is planted [19] . Furthermore, the winter wheat crop requires water to grow and be of benefit to the cotton crop. The measured daily ET rate of irrigated winter wheat varied between 1 and 2 $mm \cdot d^{-1}$ with a maximum of 13 $mm \cdot d^{-1}$ during the late spring [20] . Therefore, it was concluded that for the Texas High Plains under dryland conditions the establishment of a wheat residue is perhaps not practical and will require irrigation to be of value [19] .

The short- and long-term benefits of residues on soil physical and chemical properties, and crop production has been widely investigated, e.g., [15] [21] - [23] . Residues increase infiltration of water, from either rain or irrigation, by increasing the surface roughness and flow path tortuosity and thus increasing the time for more water to infiltrate by surface ponding [19] [24] [25] . Residues, also modify the kinetic energy from raindrops impacting the soil surface [26] [27] , which leads to less erosion [26] and increasing rain interception [28] by as much as 10% of the annual rainfall [29] . A long-term benefit of residues is to increase water storage and thus providing more water for crop production [30] [31] .

The impact of residues on E_{soil} is well documented [4] [5] [8] [32] - [34] . Residues can reduce the rate of E_{soil} in the so-called first stage [25] [30] [34] - [37] . The type of residue also affects the E_{soil} rate and wheat, for example, is more effective than cotton [35] . The presence of a residue reduces E_{soil}; however, the total evaporation equilibrates between residue covered soil and bare soil after a long period of time between water inputs [35] . For example, the rate of E_{soil} from a bare soil and a residue-covered soil will be different; however, the total amount of water lost from the bare and residue-covered soil will be very similar after a long period of time [5] [8] . The rate of E_{soil} is defined by exchanges of energy and water between the soil surface and the surrounding environment [32] and determined by the combined water and energy balance of the soil surface. The aerodynamic resistance to vapor and latent heat flux from the soil surface to the atmosphere is increased by the presence of a residue, which decreases the rate of E_{soil} [5] [8] [38] [39] . Another effect of a residue is to reduce the amount of irradiance that reaches the surface, decreasing soil temperature and E_{soil}.

The presence of a surface residue modifies the E_{soil} and affects the WUE [8] . The daily E_{soil} and crop transpiration (T) for a 100-day growing season of cotton planted into terminated wheat and conventionally tilled soil where compared by Lascano et al. [8] . In this experiment daily E_{soil} was measured for a seven-day period using microlysimeters [32] and calculated using the mechanistic ENWATBAL model [4] [40] . Both cotton cropping systems, residue and conventional, had the same seasonal ET. However, the conventional tilled cotton evaporated more E_{soil} and had a lower amount of plant T than the cotton planted into the terminated wheat. The wheat residue increased the seasonal crop T, i.e., more water used for plant growth, and reduced the seasonal E_{soil}, resulting in an increase of lint yield of 35% compared to the cotton planted in the conventional tillage with the same amount of water. The WUE of the cotton planted in the wheat residue was 2.6 g lint per kg of water used in ET, which was 27% more lint than the conventionally planted cotton [8] . In these field experiments the amount of water used to grow and establish the wheat residue were not measured and thus the calculated value of WUE is biased towards the cotton lint yield from the residue.

Our hypothesis is that despite the many benefits that residue offer, in a semiarid environment establishing the residue would use more water than conserved when used along a cotton crop. In our hypothesis the assumption was made that the cover crop that provides the residue and cotton were supplemented as needed with irrigation and each tillage treatment received the same amount of water annually. Further, it was postulated that the use of chemically terminated winter wheat residue would not significantly increase lint yields because less water would available for cotton growth and lint production. The main objective of this study was to examine and compare the water balance of two cotton cropping systems for a two-year period for conventionally tilled cotton, and for cotton planted into terminated wheat. Specifically, our objectives were to 1) compare the water used by wheat vs. soil water lost to evaporation from the conventionally tilled bare soil; 2) determine if wicking occurred after the wheat was terminated and before the cotton was planted; and 3) determine the effects that bare soils and residue- covered soils have on the water available for cotton growth and lint yield.

2. MATERIALS AND METHODS

This study was done during two growing seasons, 1994 and 1995, at experimental fields within two research farms managed by Texas Tech University, Lubbock, TX. One farm was in central Terry County, TX near Brown- field (33°10'47"N - 102°16'15"W) where the site has an Amarillo loamy fine sand soil [fine-loamy, mixed, thermic, Torrertic Paleustalf]. The other farm was in northern Lubbock County, TX near New Deal (33°44'13.76"N, 101°43'58.04"W) with a Pullman clay loam soil [fine, mixed, thermic, Torrertic Paleustoll]. These two research farms were part of a long-term (1991-2001) study to evaluate dryland cropping systems, cotton and sorghum (Sorghum bicolor L.) in the semiarid climate of the Texas High Plains. Results presented here are partially based on work by Vorheis [41] and Ralston [42] .

2.1. Irrigation

Irrigation-water at each location was applied using sprinkler irrigation methods. At Brownfield, TX a center pivot, 400 m long, irrigation system provided 50, 75, and 100% of the weekly crop water use with Low Energy Precision Application (LEPA) spray nozzles [43] . The linear (300 m in length) irrigation system used at New Deal provided 40% or 80% of weekly crop water use with LEPA spray nozzles. Half the field at New Deal was irrigated at 40% ET_0 and the other half irrigated at 80% ET_0.

At each location, the amount of water applied was based on calculated daily grass reference ET_0using as weather input net irradiance, air temperature and humidity, and wind speed measured at each site, and using the Penman-Monteith method as given by Allen et al. [44] . The daily water use was determined as ET_0 multiplied by a locally developed crop coefficient (K_c) that depended on the ground cover [45] . The weekly crop water use was calculated using the appropriate ET_0 and K_c for that site. Date and amount of water applications for Brownfield are given in Table 1 and for New Deal in Table 2. Rainfall was measured at each site using a tipping bucket. Irrigation treatments at both sites did not begin until cotton emergence. The irrigation levels were different between the two locations because the two studies were independent of each other, with different soil textures and environmental conditions.

2.2. Experimental Design

The experimental design at both locations was a randomized block, split plot setup. The experimental field at each site was about 46 ha. The major plots were tillage effects and the irrigation treatments were the split plots. All treatments were replicated four times and the plot size of each replicate was 156 ´ 183 m. The irrigation treatments (i.e., 50% and 100% ET_0 at Brownfield, and 40% and 80% ET_0 at New Deal) were started after the cotton emerged (late May through early June). Therefore, all plots for each location received identical irrigation amounts between cotton harvest from the previous season until cotton emergence the next season. One half of each irrigation treatment at each location was planted conventionally with cotton (left fallow in winter) and the other half was planted with cotton in terminated winter wheat. Locations were previously used for cotton production prior to this study with half of each field in terminated wheat residue and half of each field under conventional tillage. Treatments were replicated, n = 4, and mean separation was done using the PROC MIXED procedure and with t-type confidence interval of 0.05 using SAS on an Apple^{ò} computer (WMware and SAS version 9.2).

2.3. Cultural Practices

Winter wheat at both locations was drilled at a rate of 30 kg·ha^{-1} at 0.2-m row spacing into listed soil resulting in two rows of wheat in each furrow on 0.75-m centers. The tops of the beds were left bare in preparation for the planting of cotton. The wheat (TAM-200, Texas A&M University Foundation Seed, College Station, TX)# grew until it was tall, about 0.3 m, enough to provide wind protection, which was usually in mid- to late-April and was then terminated with the herbicide glysophate (Roundup$^{\grave{O}}$, Monsanto Company, St. Louis, MO) at a rate of 2 L·ha^{-1}. Cotton (HS-26, Paymaster Technology Corp., Scott, MS) was planted into the terminated wheat and on the fallow plots during early to mid-May. A complete fertilizer blend consisting of 50-50-0-10 was applied to the entire area after each cotton harvest and prior to planting the wheat in the fall each year. Supplemental N was provided through the irrigation water at a rate of 0.2 kg N mm^{-1}·ha^{-1} with an irrigation volume of 25 mm for the 100% and 80% irrigation treatments through the third week of flowering. Important dates for different cultural operations at both locations are given in Table 3.

2.4. Measurements

Soil water contents were measured gravimetrically at various times throughout the year, usually one to two times per month [46] , and these were converted to volumetric water content using appropriate bulk density values provided by Baumhardt et al. [47] . On each sampling date, 6 - 10 soil samples per plot were taken in 0.30-m increments to a 1.2-m depth. Water used through E_{soil} and/or T during different crop stages was based on the water balance, inputs and outputs and included precipitation, irrigation and net change in soil water content. When no crops were on the ground, all the water loss was assumed to be through E_{soil}, and if a crop did exist, the water loss was assumed to be a combination of E_{soil} and T [48] . Specifically, the following quantities were determined: 1) water required to grow an adequate amount of residue; 2) water used by the wheat after being chemically terminated; 3) water lost through E_{soil}; 4) water used by each cotton crop; and 5) the effect of each cotton cropping system on cotton lint yields. Runoff was assumed to be negligible because of slopes $< 0.1\%$ and soil water content below the 1.2-m depth was not measured because of the existence of a caliche layer around the 1.4-m depth. Cotton lint yield was estimated by hand harvesting 10 m^2 of plot area in each replication of each treatment. Cotton yield components were determined and their relative contribution to total lint yield evaluated.

Table 1. Irrigation applied and rain for two irrigation treatments and two growing seasons in Brownfield, TX.

Date	Irrigation Treatment [mm]		Rain [mm]	Date	Irrigation Treatment [mm]		Rain [mm]
	50% ET_o	100% ET_o			50% ET_o	100% ET_o	
28-Apr-94			7.4	04-Jan-95			4.6
29-Apr-94	25.4	25.4		20-Jan-95			20.1
29-Apr-94			8.4	27-Jan-95			3.3
30-Apr-94			18.0	18-Feb-95			13.0
3-May-94			1.0	01-Mar-95			1.0
10-May-94			15.5	10-Mar-95			5.6
11-May-94			29.2	07-Apr-95			8.1
12-May-94			54.1	09-Apr-95	25.4	25.4	
13-May-94			32.3	15-Apr-95	25.4	25.4	
18-May-94	6.4	6.4		02-May-95			1.3
25-May-94			3.3	05-May-95			10.9
26-May-94			1.5	06-May-95			4.1
27-May-94			5.6	11-May-95			11.4
31-May-94	25.4	25.4		15-May-95			1.5
3-Jun-94	13.2	13.2		24-May-95			1.1
10-Jun-94	25.4	25.4		30-May-95			41.0
12-Jun-94			1.5	04-Jun-95			6.4
21-Jun-94	25.4	25.4		10-Jun-95			13.5
12-Jul-94	10.8	21.6		23-Jun-95			18.0
13-Jul-94			1.8	24-Jun-95			7.0
14-Jul-94			3.8	27-Jun-95			21.0
22-Jul-94	25.4	50.8		05-Jul-95	14.0	28.0	
28-Jul-94			14.5	15-Jul-95			1.3
30-Jul-94			14.5	18-Jul-95	10.0	20.3	
3-Aug-94			3.0	20-Jul-95			5.3
8-Aug-94	25.4	50.8		21-Jul-95			2.8
27-Aug-94	12.5	25.4		25-Jul-95	10.0	20.3	
1-Sep-94			11.7	28-Jul-95	6.3	12.7	
9-Sep-94			2.0	31-Jul-95			1.3
15-Sep-94			3.8	01-Aug-95			5.1
16-Sep-94			7.9	14-Aug-95			58.0
7-Oct-94			10.7	21-Aug-95			6.0
8-Oct-94			5.3	09-Sep-95			19.3
15-Oct-94			3.6	10-Sep-95			13.7
18-Oct-94			2.3	13-Sep-95			3.0
5-Nov-94			5.8	16-Sep-95			134.0
24-Nov-94			4.3	22-Sep-95			11.0
				28-Sep-95			5.0
Total [mm]	**195.3**	**269.8**	**272.8**		**91.1**	**132.1**	**458.7**

Table 2. Irrigation applied and rain for two irrigation treatments and two growing seasons, in New Deal, TX.

Date	Irrigation Treatment [mm]		Rain [mm]	Date	Irrigation Treatment [mm]		Rain [mm]
	40% ET_o	80% ET_o			40% ET_o	80% ET_o	
26-May-94			14.0	22-Jan-95			6.9
13-Jun-94	25.4	25.4		29-Jan-95			3.0
16-Jun-94			3.0	10-Feb-95	38.1	38.1	
29-Jun-94			19.0	18-Feb-95			3.3
13-Jul-94			51.0	5-Mar-95			1.5
14-Jul-94			10.0	12-Mar-95			6.9
19-Jul-94	25.4	50.8		18-Mar-95			1.3
21-Jul-94			5.0	25-Mar-95			1.3
2-Aug-94			1.0	1-Apr-95			2.3
11-Aug-94	30.8	63.5		8-Apr-95			18.3
14-Aug-94			3.6	15-Apr-95	50.8	50.8	
8-Sep-94			7.4	16-Apr-95			8.1
14-Sep-94			11.4	18-Apr-95			1.0
15-Sep-94			8.4	5-May-95			18.5
7-Oct-94			4.3	6-May-95			5.1
14-Oct-94			1.5	15-May-95			14.5
15-Oct-94			4.6	23-May-95			2.3
18-Oct-94			1.3	24-May-95			1.8
19-Oct-94			3.3	25-May-95			2.0
20-Oct-94			1.8	26-May-95			5.1
4-Nov-94			12.7	29-May-95			16.3
19-Nov-94			2.0	30-May-95			7.9
6-Dec-94			1.0	2-Jun-95			1.0
27-Dec-94			1.3	3-Jun-95			6.4
30-Dec-94			1.0	10-Jun-95			13.5
				2-Jul-95			14.5
				14-Jul-95	16.5	33.0	
				18-Jul-95			3.3
				24-Jul-95	10.1	20.3	
				26-Jul-95	10.1	20.3	
				31-Jul-95			4.8
				1-Aug-95			2.7
				2-Aug-95			1.3
				19-Aug-95	25.4	50.8	
				9-Sep-95			2.8
				10-Sep-95			13.2
				12-Sep-95			13.5
				15-Sep-95			90.0
				19-Sep-95			55.0
				29-Sep-95			100.0
Total [mm]	**81.6**	**139.7**	**168.6**		**151.0**	**213.3**	**449.4**

Table 3. Cultural operations at two sites and growing seasons.

Cultural Practice	New Deal	Brownfield
Wheat Planted	3-Dec-93	15-Dec-93
	1-Dec-94	4-Dec-94
Wheat Terminated	4-May-94	28-Apr-94
	26-Apr-95	21-Apr-95
Cotton Planted	20-May-94	5-May-94
	15-May-95	12-May-95
Cotton Harvested	6-Nov-94	7-Nov-94
	10-Nov-95	2-Nov-94

2.5. Additional Experiment

An additional experiment was done in New Deal, TX in 1996 to measure evaporative losses of cotton planted in a bare soil and wheat terminated residue, without any irrigation. To measure the loss of water by evaporation, gravimetric soil samples were taken in the top 0.60 m of the soil profile in about five-day increments starting on the day the wheat was terminated by applying glysophate, 15 May 1996, and ending on the 3 June 1996, when more than 75% of the cotton planted had emerged. Gravimetric soil water content was converted to a volume basis using the measured soil bulk density and these measurements were replicated six times. The bare soil and wheat residue treatment was replicated three times and each plot was the same size as previously used, i.e., 156 ´ 183 m.

3. RESULTS AND DISCUSSION

3.1. Wheat Water Use

The first research objective was to measure and compare the water used by the wheat to the soil water lost to evaporation from conventionally tilled bare soil. The amount of water used to establish a wheat residue and the amount of water lost to bare E_{soil} for Brownfield and New Deal are given inTable 4. During the wheat growth stage, all plots received the same irrigation amount. These results showed that in 1995, the wheat used more ($P \leq 0.05$) water than the bare soil lost E_{soil} in both treatments and locations. At Brownfield, the wheat used 40% more water ($P \leq 0.05$) than the bare soil lost to water evaporation in the loamy fine sand textured soil. At New Deal, the wheat used 36% more water ($P \leq 0.05$) than the bare soil lost to evaporation in the clay loam soil. During the spring of 1995, establishing a wheat residue required more water compared to leaving the soil bare at both locations and on both soil textures.

Table 4. Water balance (inputs and outputs) at two locations for the 1995-growing season.

Year	Brownfield				New Deal			
	Inputs [mm]		Outputs[#] [mm]		Inputs [mm]		Outputs[#] [mm]	
			Residue	Treatment			Residue	Treatment
	Irrigation*	Rain	Bare Soil	Residue	Irrigation*	Rain	Bare Soil	Residue
1995	53.0	62.0	94.5 (17)	132.5 (16)	89.0	55.0	105 (12)	142.5 (13)

*Irrigation treatments were not active during the wheat growth, [#]Calculated values of outputs are the mean of all measurements and the standard deviations are given in parenthesis

The water depth in the upper 0.6 m of the soil profile at wheat planting and at wheat termination at both sites, Brownfield and New Deal, TX are given in Table 5. Even though all the plots within each location received the same amount of irrigation during the wheat growth period, our measurements demonstrated the variability in soil water content at each location. The coefficient of variation of the measured mean soil water content was > 25% (data not shown). This variability could be explained by the irrigation treatments applied during the previous season because overall soil water content could increase over the growing season in the highest irrigation treatments compared to the lower ones.

At both locations, the bare soil treatment resulted in an increase of the soil water content in the top 0.6 m by > 30% ($P \leq 0.05$) between the time when the wheat was terminated and planted (Table 5). However, the wheat residue treatments showed no gain in stored water at wheat termination compared to the amount of soil water at wheat planting. In the loamy fine sand, the wheat residue caused a reduction ($P \leq 0.05$) of 14% in soil water content. At Brownfield, the bare soil gained 71% ($P \leq 0.05$) soil water and at New Deal the gain was 37% ($P \leq 0.05$) from wheat planting to wheat termination. The statistical difference ($P \leq 0.05$) between the amount of soil water at wheat termination at Brownfield was 36 mm between the bare soil and wheat residue treatments. The statistical difference ($P \leq 0.05$) between soil water at wheat termination at New Deal was 40 mm of water in the top 0.6 m of soil. Therefore, there was no gain, 36 vs. 40 mm, in stored water by the wheat residue plots over the spring of 1995 at either location, while the bare soil was able to increase the water content in the top 0.6 m of soil by more than 30% ($P \leq 0.05$) over the spring months.

3.2. Wicking

The second objective was to determine if upward movement of water through the wheat residue, i.e., wicking, occurred after the wheat was terminated and before the cotton was planted. The amount of water lost to evaporation from both the bare soil and terminated wheat plots between wheat termination and cotton emergence for 1995 at Brownfield and New Deal are shown as a bar graph in Figure 1. These results suggested, and as expected the evaporation of water was greater in the finer textured soil when compared to the coarser textured soil. In the fine loamy sand soil, the bare soil evaporated 44% more ($P \leq 0.05$) soil water than the wheat residue. However, in the clay loam soil the wheat residue evaporated 66% more ($P \leq 0.05$) soil water than the bare soil. This result suggested that there might be a soil textural effect on the amount of water that can evaporate from a bare soil and from a terminated wheat residue [19] .

It was speculated that differences in water evaporation from the bare and from the wheat-covered soil were due to the different textures and the process of wicking. The soil at New Deal is a clay loam, i.e., more conducive for upward capillary flow of water; while Brownfield has a loamy fine sand, whose large pores would decrease capillary flow [13] . Under wicking, wheat residue could lose more water to evaporation than bare soil because many roots act as capillaries and the surface area provided by the leaves of the terminated wheat contribute to the evaporation. A possible solution to this problem could be to mechanically sever the roots of the wheat plant after termination at a 0.2 - 0.3 m depth and in this way disrupt the capillary movement of water from the soil to the atmosphere.

The cumulative evaporation of soil water from the bare soil and terminated wheat, between wheat termination to cotton emergence is shown in Figure 2. The bare soil had an 18% larger ($P \leq 0.05$) amount of water evaporation loss compared to the wheat residue. This is a result that in part may be explained by the frequent irrigation and rain events that occurred during the study. These wetting events resulted in a bare soil that was wetter than the top of the wheat residue, leading to near free-water evaporation to occur at faster rate than wicking. However, the results from this study were not conclusive to determine the effects of wicking in a clay loam soil and the measurement of E_{soil}.

Table 5. Amount of water in the top 0.6 m of the soil for wheat residue at planting and termination at two locations.

Year	Brownfield Residue Treatment [mm]				New Deal Residue Treatment [mm]			
	Bare Soil		Residue		Bare Soil		Residue	
	Planting	Termination	Planting	Termination	Planting	Termination	Planting	Termination
1995	47.0	80.5	51.5	44.5	104.0	142.5	102.0	103.0

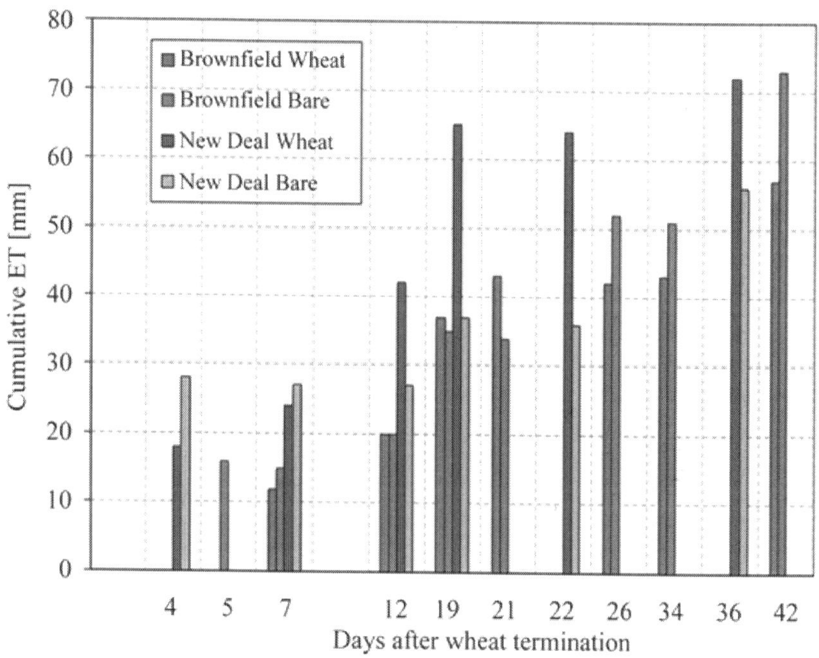

Figure 1. Cumulative measured evapotranspiration (ET, mm) as a function of days after wheat termination at Brownfield (BF) and New Deal (ND) from the wheat and bare soil residue treatment for the 1995- growing season.

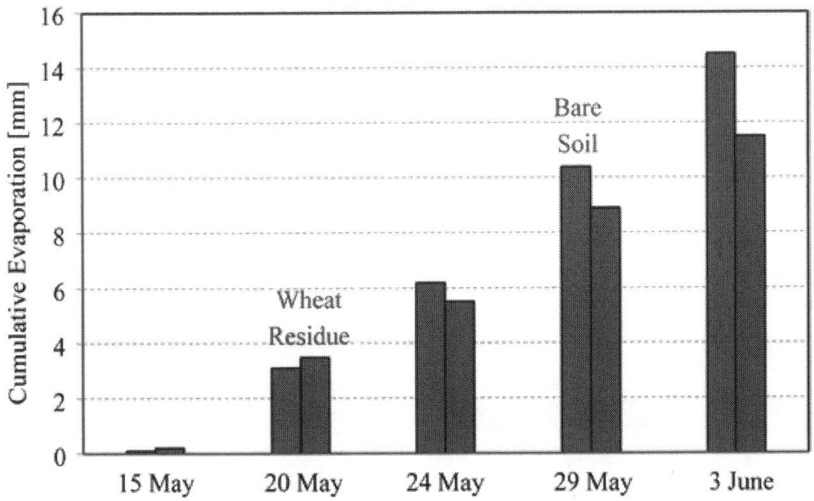

Figure 2. Cumulative measured evaporation in mm from a bare soiland wheat residue treatment for the time period between wheat termination (15 May 1996) to cotton emergence (3 June 1996) in New Deal TX.

Part of the difficulty in determining differences of E_{soil} under field conditions is that the amount of water that may be lost due to wicking is of the order of 1 mm·d^{-1} and this loss takes place from mainly the surface 0.1 m layer of the soil profile. This rate of water evaporation represents a change of 0.01-m^3·m^{-3} volumetric water content on a daily basis, which is beyond the sensitivity of current instrumentation to measure volumetric soil water content [46]. Nevertheless, a wicking loss of 1 mm·d^{-1} for 20 days represents 20 mm of water, and this amount of water could determine the difference of having adequate water in the seedbed to germinate and to emerge the cotton crop planted in the residue [13].

The water used by the wheat and lost by E_{soil} affects the amount of water available for cotton growth. The amount of water content in the 1.2 m soil profile at cotton planting for both sites and growing seasons season is shown in Table 6. The bare soil treatments had more water at cotton planting than the wheat residue treatments at both locations except for the 40% ET_o plots at New Deal and the 50% ET_o plots at Brownfield for 1994. All the treatments in New Deal excluding the 40% ET_o irrigation treatment in 1994 showed a significant increase ($P \leq 0.05$) in soil water for the bare soil treatments compared to the wheat residue treatments, while Brownfield did not have a significant increase ($P \leq 0.05$) in soil water for the bare soil treatments at cotton planting. This stored water is important, particularly at the time when the cotton is planted as it necessary for seed germination and emergence impacting the vitality and vigor of the cotton seedling, which can later affect lint yields [49] [50] .

Even though the wheat used more water in Brownfield in 1995 than the bare soil (Table 5), it appears that the wheat only extracted water from the top 0.6 m of soil because the water depth in the top 1.2 m of soil at cotton planting (Table 6) showed that the soil profile for each residue treatment contained about the same amount of water in the top 1.2 m of soil. However, at New Deal, capillary movement of water by a hydraulic gradient and by wheat roots may have caused a reduction in the amount of water in the top 1.2 m of soil for the wheat residue plots, allowing the bare soil to store more water at the time that cotton was planted (Table 6).

3.3. Cotton Water Use

The last and third objective of this study was to determine the effects bare and residue-covered soils have on the water available for cotton growth and development. Water inputs and water used by the cotton in both seasons and at both sites are given in Table 7. There were no significant differences ($P \leq 0.05$) in the amount of water used by the cotton for any of the residue treatments at Brownfield (Table 7). However, the 80% ET_0 bare soil treatments used at least 10% more ($P \leq 0.05$) water to grow the cotton than the residue treatments for both years at New Deal (Table 7) and as previously discussed, could be accounted for sampling error (Table 5).

Cotton lint yields were used to evaluate the effects residues have on cotton growth and development. Cotton lint yields for both locations and growing seasons are given in Table 8. These cotton lint yield values were similar to cotton yields reported for surrounding areas [51] [52] . In 1994, both locations had significantly higher ($P \leq 0.05$) lint yields (at least 8% higher) from the conventionally tilled treatments than the wheat residue plots for all irrigation treatments (Table 8). The only treatment with a loss in lint yield from the conventionally tilled treatments was in the in the low irrigation treatment at Brownfield in 1995 with a 14% decrease ($P \leq 0.05$) in lint yields. The greatest gain in lint yield from the conventionally tilled treatments over the terminated wheat treatments occurred in the 40% irrigation treatment at New Deal in 1994 with a 42% gain ($P \leq 0.05$). Therefore, our results suggested that the presence of residues had a significant ($P \leq 0.05$) impact with a loss of lint yield for both years of the study. This result confirms our hypothesis that in a semiarid region the use of a small grain cover crop under irrigated conditions uses more water than it conserves compared to conventional tilled cotton, reducing cotton lint yields.

As expected and in climates where water is a limiting factor in crop production, the irrigation treatments at each location that received the greatest amount of water resulted in higher lint yields compared to the lower irrigation amounts (Table 8). For example, Bordovsky et al. [45] showed that higher irrigation amounts also gave higher yields, but also reported that at a 50% ET_0 irrigation level, conventionally tilled and wheat residue cotton systems yielded about the same ($P \leq 0.05$), while at a 100% ET_0 irrigation level, the conventionally tilled cotton cropping system yielded 17% less ($P \leq 0.05$) as compared to the wheat residue treatment at Lamesa, TX, which has a sandy loam soil.

Table 6. Amount of water in the top 1.2 m of the soil when cotton was planted at two locations and two years.

Year	Brownfield Residue Treatment [mm]		New Deal Residue Treatment [mm]	
	Bare Soil	Residue	Bare Soil	Residue
1994	207.0	201.5	296.0	299.0
1995	177.5	169.5	238.0	192.0

Table 7. Water balance (inputs and outputs), for two irrigation treatments at two locations and two years.

Brownfield							
Year	Water Inputs [mm] Irrigation Treatment		Rain [mm]	Water Outputs [mm] Irrigation Treatment			
				$50\%\ ET_0$ Residue Treatment		$100\%\ ET_0$ Residue Treatment	
	$50\%\ ET_0$	$100\%\ ET_0$		Bare Soil	Residue	Bare Soil	Residue
1994	125	200	93	276	295	395	403
1995	41	81	375	315	314	406	403

New Deal							
Year	Water Inputs [mm] Irrigation Treatment		Rain [mm]	Water Outputs [mm] Irrigation Treatment			
				$40\%\ ET_0$ Residue Treatment		$80\%\ ET_0$ Residue Treatment	
	$40\%\ ET_0$	$80\%\ ET_0$		Bare Soil	Residue	Bare Soil	Residue
1994	83	140	172	283	304	442	387
1995	93	130	382	541	539	620	558

Table 8. Average cotton lint yields for two locations, and two irrigation and residue treatments, for two years.

Brownfield – Cotton Lint Yield [kg·ha⁻¹]				New Deal Cotton Lint Yield [kg·ha⁻¹]			
Irrigation Treatment				Irrigation Treatment			
50% ET_o		100% ET_o		40% ET_o		80% ET_o	
Residue Treatment		Residue Treatment		Residue Treatment		Residue Treatment	
Bare Soil	Residue	Bare Soil	Residue	Bare Soil	Residue	Bare Soil	Residue
440 (20)*	392 (16)	933 (31)	816 (11)	561 (19)	394 (23)	674 (11)	599 (21)
380 (17)	444 (21)	594 (26)	546 (24)	436 (11)	401 (31)	603 (19)	559 (11)

*The standard error of the mean is in parenthesis. Mean separation was calculated using PROC MIXED with a t-type confidence interval of 0.05

4. CONCLUSIONS

In this study, the cotton water use and lint yield of two cotton-cropping systems in the semiarid climate of the Texas High Plains was compared. These cropping systems were conventionally tilled and cotton planted into terminated wheat, on two sites and a low and high irrigation rate over two growing seasons. Our objective was to evaluate the effectiveness of planting cotton into a crop residue obtained by terminating a winter wheat crop planted in December. The purpose of the residue is mainly to provide a mechanical barrier to wind and thus protect cotton seedlings at a time when they are most vulnerable to sandblast damage due to high wind speeds. The selected sites, allowed us to compare the impact of soil texture, i.e., coarse (Brownfield, TX) and fine (New Deal, TX). The residue treatments did not affect the total amount of water evaporated and transpired during the two seasons. However, at both locations the bare residue treatment had a higher lint yield than the cotton grown under the wheat residue. Wicking, the capillary movement of water through the wheat residue was only statistically significant at the site with the coarser texture (Brownfield, TX) and only in one (1994) of the two years.

These results suggest that the use of terminated wheat residue did not have a positive impact on increasing water storage for the subsequent cotton crop. Furthermore, terminated wheat residue adversely affected cotton lint yield on both years and sites and thus no advantages found in planting cotton into a terminated wheat residue in the semiarid Texas High Plains. Nevertheless, over a longer period of time, the use of a wheat residue may improve soil structure, which may lead to higher infiltration rate reducing runoff and storing more rainfall.

It was postulated that in a semiarid climate growing a small grain cover crop might use more water than it conserves when combined with a cotton crop. During the 1995 wheat growth period, the conservation tillage treatment used 20% more water than the conventional tillage treatment for all treatments at both locations supporting our hypothesis. It was further hypothesized that the use of residue would not increase cotton lint yields, and all the conventionally tilled treatments for two seasons had higher lint yields except for one plot in 1994 at Brownfield over the wheat residue plots. Therefore, for these two locations for the 1994 and 1995 cotton crop seasons, the use of terminated wheat residue did not improve the growing conditions and environment for the cotton crops compared to conventional tillage cropping systems. The significance of this work is that growers using this "conservation compliance" tillage system may not be using water optimally and may even be experiencing decreased cotton lint yields.

ACKNOWLEDGEMENTS

This work is a tribute to the many research contributions of Dr. Dan Krieg, Professor emeritus, Plant and Soil Science, Texas Tech University, to crop physiology and agronomy of dryland cropping systems. On the 18 June 2015 he lost his battle with cancer.

REFERENCES

1. USDA-NASS (2014) Texas Crop Production, USDA—National Agricultural Statistics Service, Issue No. PR-163-14, 10 December 2014.
2. Lascano, R.J. (2000) A General System to Measure and to Calculate Daily Crop Water Use. Agronomy Journal, 92, 821-832.
3. Colaizzi, P.D., Gowda, P.H., Marek, T.H. and Porter, D.O. (2009) Irrigation in the Texas High Plains: A Brief History and Potential Reductions in Demand. Irrigation & Drainage, 58, 257-274.
4. Lascano, R.J., Van Bavel, C.H.M., Hatfield, J.L. and Upchurch, D.R. (1987) Energy and Water Balance of a Sparse Crop: Simulated and Measured Soil and Crop Evaporation. Soil Science Society of America Journal, 51, 1113-1121.
5. Lascano, R.J. and Baumhardt, R.L. (1996) Effects of Crop Residue on Soil and Plant Water Evaporation in a Dryland Cotton System. Theoretical Applied Climatology, 54, 69-84.

6. Krieg, D.R. (1996) Physiological Aspects of Ultra Narrow Row Cotton Production. Proceedings of the Beltwide Cotton Conference, Nashville, TN. National Cotton Council, Memphis, 9-12 January 1996, 66.

7. Darawsheh, M.K., Khah, E.M., Aivalakis, G., Chachalis, D. and Sallaku, F. (2009) Cotton Row Spacing and Plant Density Cropping Systems. I. Effects on Accumulation and Partitioning of Dry Mass and LAI. Journal of Food, Agriculture & Environment, 7, 258-261.

8. Lascano, R.J., Baumhardt R.L., Hicks, S.K. and Heilman, J.L. (1994) Soil and Plant Water Evaporation from Strip-Tilled Cotton: Measurement and Simulation. Agronomy Journal, 86, 987-994.

9. Musick, J.T., Pringle, F.B., Harman, W.L. and Stewart B.A. (1990) Long-Term Irrigation Trends-Texas High Plains. Applied Engineering Agriculture, 6, 717-724.

10. Lascano, R.J. and Nelson, J.R. (2014) Circular Planting to Enhance Rainfall Capture in Dryland Cropping Systems at a Landscape Scale: Measurement and Simulation. In: Ahuja, L.R., Ma, L. and Lascano, R.J., Eds., Practical Applications of Agricultural System Models to Optimize the Use of Limited Water, Advances in Agricultural Systems Modeling, ASA, CSSA, SSSA, Madison, Volume 5, 85-111.

11. Bordovsky, J.P., Lyle, W.M. and Keeling, J.W. (1994) Crop Rotation and Tillage Effects on Soil Water and Cotton Yield. Agronomy Journal, 86, 1-6.

12. Baker, J.T., McMichael, B., Burke, J.J., Gitz, D.C., Lascano, R.J. and Eprath, J.E. (2009) Sand Abrasion Injury and Biomass Partitioning in Cotton Seedlings. Agronomy Journal, 101, 1297-1303.

13. Koekkoek, E.J.W., Lascano, R.J., Hicks, S.K., Krieg, D.R. and Stroosnijder, L. (1995) Loss of Water through Terminated Wheat Plants: A Wick Effect. Proceedings of the 1995 Annual Meetings, St. Louis, 29 October-3 November 1995, 15.

14. Van Rensburg, L.D. (2010) Advances in Soil Physics: Application in Irrigation and Dryland Crop Production, South African. Journal of Plant and Soil, 27, 9-18.

15. Unger, P.W. (1994) Residue Management for Winter Wheat and Grain Sorghum Production with Limited Irrigation. Soil Science Society American Journal, 58, 537-542.

16. Nielsen, D.C., Unger, P.W. and Miller, P.R. (2005) Efficient Water Use in Dryland Cropping Systems in the Great Plains. Agronomy Journal, 97, 364-372.

17. AG-CARES (2004-2008) AG-CARES Annual Report. Texas A & M AgriLife Research & Extension Center, Lubbock.

18. Helms Farm Research Reports (2001-2005) Helms Farm Annual Reports. Texas A & M AgriLife Research & Extension Center, Lubbock.

19. Jones, O.R., Hauser, V.L. and Popham, T.W. (1994) No-Tillage Effects on Infiltration, Runoff, and Water Conservation on Dryland. Transactions of the ASAE, 37, 473-479.

20. Howell, T.A., Steiner, J.L., Schneider, A.D. and Evett, S.R. (1995) Evapotranspiration of Irrigated Winter Wheat— Southern High Plains. Transactions of the ASAE, 38, 745-759.

21. Guerif, J., Richard, G., Durr, C., Machet, J.M., Recous, S. and Roger-Estrade, J. (2001) A Review of Tillage Effects on Crop Residue Management, Seedbed Conditions and Seedling Establishment. Soil & Tillage Research, 61, 13-22.

22. Hobbs, P.R., Sayre, K. and Gupta, R. (2008) The Role of Conservation Agriculture in Sustainable Agriculture. Philosophical Transactions of the Royal Society B: Biological Sciences, 363, 543-555.

23. Alvarez, R. and Steinbach, H.S. (2009) A Review of the Effects of Tillage Systems on Some Soil Physical Properties, Water Content, Nitrate Availability and Crops Yield in the Argentine Pampas. Soil & Tillage Research, 104, 1-15.

24. Baumhardt, R.L., Keeling, J.W. and Wendt, C.W. (1993) Tillage and Residue Effects on Infiltration into Soils Cropped to Cotton. Agronomy Journal, 85, 379-383.

25. Dao, T.H. (1993) Tillage and Winter Wheat Residue Management Effects on Water Infiltration and Storage. Soil Science Society of America Journal, 57, 1586-1595.

26. Hoogmoed, W.B. and Stroosnijder, L. (1984) Crust Formation on Sandy Soils in the Sahel. I. Rainfall and Infiltration. Soil & Tillage Research, 4, 5-23.

27. Lascano, R.J., Vorheis, J.T., Baumhardt, R.L. and Salisbury, D.R. (1997) Computer-Controlled Variable Intensity Rain Simulator. Soil Science Society of America Journal, 61, 1182-1189.

28. Alberts, E.E. and Neibling, W.H. (1994) Influence of Crop Residues on Water Erosion. In: Unger, P.W., Ed., Managing Agricultural Residues, Lewis Publ., Chelsea, MI, 19-39.

29. Savabi, M.R. and Stott, D.E. (1994) Plant Residue Impact on Rain Interception. Transactions of the ASAE, 37, 1093-1098.

30. Steiner, J.L. (1994) Crop Residue Effects on Water Conservation. In: Unger, P.W., Ed., Managing Agricultural Residues, Lewis Publ., Chelsea, MI, 41-76.

31. Baumhardt, R.L., Schwartz, R., Howell, T.A., Evett, S.R. and Colaizzi, P. (2013) Residue Management Effects on Water Use and Yield of Deficit Irrigated Cotton. Agronomy Journal, 105, 1026-1034.

32. Lascano, R.J. and Van Bavel, C.H.M. (1986) Simulation and Measurement of Evaporation from a Bare Soil. Soil Science Society of America Journal, 50, 1127-1132.

33. Farahani, H.J. and Ahuja, L.R. (1996) Evapotranspiration Modeling of Partial Canopy/Residue-Covered Fields. Transactions of the ASAE, 39, 2051-2064.

34. Klocke, N.L., Currie, R.S. and Aiken, R.M. (2009) Soil Water Evaporation and Crop Residues. Transactions of the ASABE, 52, 103-110.

35. Steiner, J.L. (1989) Tillage and Surface Residue Effects on Evaporation from Soils. Soil Science Society American Journal, 53, 911-916.

36. Enz, J.W., Brun, L.J. and Larsen, J.K. (1988) Evaporation and Energy Balance for Bare and Stubble Covered Soil. Agricultural and Forest Meteorology, 43, 59-70.

37. Horton, R., Kluitenberg, G.J. and Bristow, K.L. (1994) Surface Crop Residue Effects on the Soil Surface Energy Balance. In: Unger, P.W., Ed., Managing Agricultural Residues, Lewis Publ., Chelsea, MI, 143-162.

38. Heilman, J.L., McInnes, K.J., Gesch, R.W. and Lascano, R.J. (1992) Evaporation from Ridge-Tilled Soil Covered with Herbicide-Killed Winter Wheat. Soil Science Society American Journal, 56, 1278-1286.

39. Fryrear, D.W. and Bilbro, J.D. (1994) Wind Erosion Control with Residues and Related Practices. In: Unger, P.W., Ed., Managing Agricultural Residues, Lewis Publ., Chelsea, MI, 7-17.

40. Evett, S.R. and Lascano, R.J. (1993) ENWATBAL.BAS: A Mechanistic Evapotranspiration Model Written in Compiled Basic. Agronomy Journal, 85, 763-772.

41. Vorheis, J.T. (1997) Water Balance of Cotton Cropping Systems. Master's Thesis, Plant and Soil Science Department, Texas Tech University, Lubbock, 38 p.

42. Ralston, J.T. (1997) Management Strategies for Dryland Cotton Production in West Texas. Master's Thesis, Plant and Soil Science Department, Texas Tech University, Lubbock, 47 p.

43. Lyle, W.M. and Bordovsky, J.P. (1981) Low Energy Precision Application (LEPA) Irrigation System. Transactions of the ASAE, 24, 1241-1245.

44. Allen, R.G., Walter, I.A., Elliot, R., Howell, T.A., Itenfisu, D. and Jensen, M.E. (2005) The ASCE Standardized Reference Evapotranspiration Equation. ASCE-EWRI Task Committee Report, January 2005, 70 p.

45. Bordovsky, J.P., Lyle, W.M., Lascano, R.J. and Upchurch, D.R. (1992) Cotton Irrigation Management with LEPA Systems. Transactions of the ASAE, 35, 879-884.

46. Evett, S.R. (2007) Soil Water and Monitoring Technology. In: Lascano, R.J. and Sojka, R.E., Eds., Irrigation of Agricultural Crops, 2nd Edition, ASA, CSSA, SSSA, Madison, 25-84.

47. Baumhardt, R.L., Lascano, R.J. and Krieg, D.R. (1995) Physical and Hydraulic Properties of a Pullman and Amarillo Soil on the Texas South Plains. Technical Report D.R. No. 95-1, Texas A & M University Agricultural Research and Extension Center, Lubbock/Halfway.

48. Bertuzzi, P., Bruckler, L., Bay, D. and Chanzy, A. (1994) Sampling Strategies for Soil Water Content to Estimate evapotranspiration. Irrigation Science, 14, 105-115.

49. Peng, S., Krieg, D.R. and Hicks, S.K. (1989) Cotton Lint Yield Response to Accumulated Heat Units and Soil Water Supply. Field Crop Research, 19, 253-262.

50. Morrow, M.R. and Krieg, D.R. (1990) Cotton Management Strategies for a Short-Growing Season Environment: Water-Nitrogen Considerations. Agronomy Journal, 92, 52-56.

51. USDA-NASS (1994) Agricultural Statistics-1994.

52. USDA-NASS (1995) Agricultural Statistics-1995. Chapter II—Statistics of Cotton, Tobacco, Sugar Crops, and Honey.

11

CHAPTER

Irrigation Water Quality Standards for Indirect Wastewater Reuse in Agriculture: A Contribution toward Sustainable Wastewater Reuse in South Korea

Hanseok Jeong [1], Hakkwan Kim [1,*] and Taeil Jang [2]

[1]*Institute of Green Bio Science and Technology, Seoul National University, Gangwon 25354, Korea*

[2]*Department of Rural Construction Engineering, Chonbuk National University, Jeonbuk 54896, Korea*

ABSTRACT

Climate change and the subsequent change in agricultural conditions increase the vulnerability of agricultural water use. Wastewater reuse is a common practice around the globe and is considered as an alternative water resource in a changing agricultural environment. Due to rapid urbanization, indirect wastewater reuse, which is the type of agricultural wastewater reuse that is predominantly practiced, will increase, and this can cause issues of unplanned reuse. Therefore, water quality standards are needed for the safe and sustainable practice of indirect wastewater reuse in agriculture. In this study, irrigation water quality criteria for wastewater reuse were discussed, and the standards and guidelines of various countries and organizations were reviewed to suggest preliminary standards for indirect wastewater reuse in South Korea. The proposed standards adopted a probabilistic consideration of practicality and classified the use of irrigation water into two categories: upland and rice paddy. The standards suggest guidelines for *E. coli*, electric conductivity (EC), turbidity, suspended solids (SS), biochemical oxygen demand (BOD), pH, odor, and trace elements. Through proposing the standards, this study attempts to combine features of both the conservative and liberal approaches, which in turn could suggest a new and sustainable practice of agricultural wastewater reuse.

Keywords: indirect wastewater reuse; irrigation water quality; paddy irrigation; wastewater reuse; water quality standards

1. INTRODUCTION

The lack of freshwater resources due to population growth and the degradation of water quality is becoming a big challenge for agricultural water [1,2]. Severe droughts due to climate change and the increase in protected cultivation are making it difficult to provide a stable supply of agricultural water [3,4]. As an alternative water resource, wastewater reuse for agriculture is gaining international interest [5,6,7]. Treated wastewater, which in many ways is ideal for agricultural water [8,9,10], is already being widely used throughout the world [11], and previous studies suggest that more than 10% of the world's population is consuming agricultural products cultivated by wastewater irrigation [12].

Agricultural wastewater reuse can be classified into direct and indirect wastewater reuses [13]. Direct wastewater reuse refers to the method whereby irrigation water is supplied directly from the wastewater treatment plant (WWTP), while indirect wastewater reuse is a method where effluent from the WWTP or untreated wastewater is collected downstream. In the case of direct wastewater reuse, the irrigation water quality is determined by the quality of WWTP effluent. Therefore, the target water quality can be easily achieved by controlling the quality of the effluent. In the case of indirect wastewater reuse, however, controlling the irrigation water quality is difficult because it is dependent on various factors such as treated wastewater effluent quality and the hydrological conditions of the stream which the effluent flows into [14]. The increase in the number of WWTPs due to rapid urbanization is intensifying the effect of indirect wastewater reuse on irrigation water [14].

When using treated wastewater for irrigation, water quality must be strictly controlled considering factors such as the possible accumulation of substances which are harmful for crop growth, the potential damage to soil by the transformation of its physical and chemical characteristics

[15,16], and microbe infection [17,18]. Using diluted treated wastewater could somewhat mitigate the side effects but nonetheless cannot prevent it from damaging the commercial value of crops. Therefore, without additional treatment, it is suggested that diluted treated wastewater be used restrictively [19]. In addition, when using treated wastewater for rice paddy irrigation, the water showed a significant level of salinity compared to conventional irrigation water, and total coliform concentration surpassed the levels of the water quality standards for direct wastewater reuse [20,21]. Previous studies suggest that the quality of irrigation water from streams fails to meet the agricultural water quality standards due to the effects of WWTP effluent [14,22]; thus, the water quality needs to be managed to ensure safe and sound crop production.

Angelakis *et al.* [23], Brissaud [24], Kalavrouziotis *et al.* [25], and Paranychianakis *et al.* [26] researched the current status of wastewater reuse for agriculture, assessed its standards and guidelines for water quality, and asserted the necessity for a common criterion based on scientific evidence. Kretschmer *et al.*[27] examined the standardization of water quality for agricultural wastewater reuse in terms of technology, economy, and human safety. Jimenez [28] studied practical criteria and treatment technologies for safe wastewater reuse in Mexico, where a significant amount of wastewater is used for agriculture. Organizations such as the Word Health Organization (WHO) [12,29] and the United States Environmental Protection Agency (US EPA) [30,31,32,33], and many countries such as Cyprus [34], France [26], Greece [35], Israel [36], Italy [34], Portugal [37], and Spain [38], have all recently suggested and modified water quality guidelines or standards for safe wastewater reuse. South Korea also laid out guidelines for irrigation water quality in 2005, and proclaimed irrigation water quality standards for direct wastewater reuse in 2011.

Most of the currently suggested water quality standards are based on the direct wastewater reuse rather than the indirect wastewater reuse, which is a predominant manner in agricultural wastewater reuse [39]. They have also focused on the upland irrigation including vegetables and have not well considered the characteristics of irrigation water in paddy fields: Irrigation water for paddy rice production accounts for more than 70% of the total irrigation water in Asia [40], and paddy rice is the largest consumer of freshwater resources in South and Southeast Asia [41,42]. This study, therefore, examines existing water quality criteria for irrigation water and wastewater reuse, and analyzes the water quality standards of other countries to set agricultural water quality standards for indirect wastewater reuse considering both paddy and upland irrigation.

2. CONSIDERATIONS OF WATER QUALITY STANDARDS FOR INDIRECT WASTEWATER REUSE

2.1. Irrigation Water Quality Criteria

In the case of indirect wastewater reuse, the irrigation water quality is affected by water with which the treated wastewater is diluted such as stream water or lake water. Therefore, to examine the water quality standards for indirect wastewater reuse, irrigation water quality criteria should be considered.

2.1.1. Salinity

Salinity has been deemed as the most important factor of agricultural water quality because high salinity in soil can create a hostile environment for the crop to absorb nutrients and provoke specific ion toxicity [12,43,44]. The Food and Agriculture Organization (FAO) has established guidelines for agricultural water primarily based on salinity. These guidelines have been modified by Ayers and Westcot in 1985 [45] and are being internationally accepted.

2.1.2. Nutrients

Nitrogen (N) and phosphorus (P) are the primary nutrients for crop growth but nonetheless when applied excessively can give a negative effect. Especially for paddy rice, excessive N and P can cause the crop to over-grow which leads to lodging [46]. This is why Taiwan and Japan, which have farming conditions similar to South Korea, limit the total nitrogen (T-N) content for paddy irrigation water [47,48]. Excessive nutrients in water can also cause groundwater contamination as well as eutrophication in coastal areas or lakes [49,50] and this is why, in South Korea, N and P contents are included in agricultural water quality standards.

2.1.3. Organic Matters

Generally, biochemical oxygen demand (BOD) is used as an index for organic matter. In a high BOD environment, oxygen in water is consumed for decomposing organic matters to create an anaerobic state, and, during the process of decomposition, oxides in the soil such as Fe^{3+}, Mn^{5+}, and SO_4^{2-}consume oxygen to lower the oxidation-reduction potential [51]. In the end, the generated iron, manganese, and sulfide along with organic acids can disrupt the paddy rice to absorb nutrients [52].

2.1.4. Hydrogen Ion Concentration

Hydrogen ion concentration represented by pH is an index upon which irrigation water is quickly assessed for its suitability. Normally, the pH of irrigation water ranges from 6.5 to 8.4. The pH outside of the normal range might be suitable for irrigating, but has the potential to cause an imbalance of nutrients or contain poisonous ions [45]. The biggest hazard related to an abnormal pH in water is its effect on irrigation facilities. Exceptionally low pH in irrigation water can expedite the corrosion process of facilities, and irrigation water containing high levels of alkalinity can lower the efficiency of the trickle irrigation system [44].

2.1.5. Trace Elements

Trace elements are necessary for crop growth but when the amount of heavy metals in irrigation water is excessive, it can cause harm. Copper (Cu) can cause leaf chlorosis as well as the suppression of root growth [52]. Zinc (Zn) and arsenic (As) both have side effects of stem chlorosis and root growth suppression [52]. Aluminum (Al), in acid soil, can decrease productivity [52]. Lead (Pb), cyan (CN), and cadmium (Cd) are normally strictly restricted since, when dissolved in water or soil, they can be accumulated in the crop and in turn harmful to the human body [53]. The FAO's irrigation water quality guidelines, therefore, has set limits on concentrations of trace elements in irrigation water [45].

2.2. Conditions of Water Quality for Wastewater Reuse

Wastewater irrigation should consider human health risk along with the conventional irrigation water quality standards. The water quality standards for agricultural wastewater reuse can be classified into the conservative standards of California and the liberal standards of the WHO. This classification is well reflected in the microbe-related categories [24]. The WHO, in 1989 and 2006, suggested health-based targets for safe agricultural wastewater reuse. The WHO limits the permitted wastewater to domestic and/or urban wastewater and does not recommend using wastewater that contains excessive amount of industrial wastewater. Using untreated wastewater for agriculture can cause parasitosis or water-borne epidemics, and there have been numerous reports on the break of infectious diseases due to inappropriate wastewater reuse [29].

In the United States (US), the EPA suggests the water quality guidelines for wastewater reuse and each state government sets their own water quality standards based on these guidelines. Compared to the WHO guidelines, the US applies strict standards, and each state's microbe standard of irrigation water and treatment requirement are shown in Table 1. Texas and Florida prohibit the direct contact of irrigation water. Although California, which set the very first regulations for using wastewater for irrigation in 1918 [54], does not list specific prohibition

Table 1. Water quality standards of bacterial indicators for food and non-food crop irrigation in United States (modified from US EPA [33]).

Classification		California	Florida [1]	New Jersey	North Carolina	Texas [1]	Virginia [2]
Food [(a)]	Bacterial indicators (cfu/100 mL)	Total coliform: −2.2 (7 days median) −23 (not more than one sample exceeds this value in 30 days) −240 (max)	Fecal coliform: −75% of samples below detection −25 (max)	Fecal coliform: −2.2 (weak median) −14 (max)	Fecal coliform or *E. coli*: −3 (monthly mean) −25 (monthly mean)	Fecal coliform or *E. coli*: −20 (30-day geom) −75 (max)	Fecal coliform: −14 (monthly geom), CAT > 49*E. coli*: −11 (monthly geom), CAT > 35
	Treatment requirements	Oxidized, coagulated, filtered, disinfected	Secondary treatment, filtration, high-level disinfection	Filtration, high-level disinfection	Filtration, dual UV/chlorination (or equivalent)	NS	Secondary treatment, filtration, high-level disinfection
Non-food [(b)]	Bacterial indicators (cfu/100 mL)	NS	Fecal coliform: −200 (avg) −800 (max)	Fecal coliform: −200 (monthly geom) −400 (weak geom)	Fecal coliform or *E. coli*: −14 (monthly mean) −25 (daily max)	Fecal coliform or *E. coli*: −200 (30-day geom) −800 (max)	Fecal coliform: −200 (monthly geom), CAT > 800*E. coli*: −126 (monthly geom), CAT > 235
	Treatment requirements	Oxidized	Secondary treatment, basic disinfection	Case-by-case	Filtration (or equivalent)	NS	Secondary treatment, disinfection

NS = not specified by the state's reuse regulation; NP = not permitted by the state; CAT = corrective action threshold; geom = geometric mean. [1] In Florida and Texas, spray irrigation (*i.e.*, direct contact) is not permitted on foods that may be consumed raw, and only irrigation types that avoid reclaimed water contact with edible portions of food crops. [2] The requirements presented for Virginia are for food crops eaten raw. [(a)] Food crops: The use of reclaimed water to irrigate food crops that are intended for human consumption. [(b)] Non-food crops and processed food crops: The use of reclaimed water to irrigate crops that are either processed before human consumption or not consumed by humans.

laws, the water quality for wastewater is managed by strict standards similar to those for drinking water.

3. GUIDELINES AND STANDARDS FOR AGRICULTURAL WASTEWATER REUSE

Water quality of agricultural wastewater reuse is presented in criteria, guidelines, and standards. Water quality criteria are the result of scientific examinations on the suitability of water to be used for certain purposes. Kretschmer *et al.* [27] and Tsagarakis *et al.* [55] have suggested water quality criteria for safe wastewater reuse. Water quality guidelines refer to a set of management targets based on the water quality criteria, the following of which is recommended but nevertheless not restricted by law. Organizations and countries (e.g., WHO [12], US EPA [33], and Australia [56]) have recommended guidelines for safe reuse of wastewater as agricultural water. In terms of water quality standards, which are the actual regulations restricted by law, Greece and the US set state-based standards, respectively [33,35]. In South Korea, the Ministry of Environment (MOE) [57] has proclaimed water quality standards for treated wastewater based on its specific purpose of use. In this study, the water quality standards of countries (Cyprus, France, Greece, Israel, Italy, Portugal, Spain, and South Korea), which have recently revised their standards and widely use treated wastewater for agriculture, as well as the guidelines of WHO [12] and US EPA [33], two major influences on water reuse criteria worldwide [23,24,26], are examined (Table 2).

3.1. WHO

The WHO first devised guidelines for wastewater reuse for irrigation in 1973, and this has become the international standard [12,29]. Many European and South American countries have adopted the WHO guidelines [29] and modified it to correspond to the geographical, epidemiological idiosyncrasies, as well as the economic conditions, of each country. However, there have been controversies regarding the fact that the WHO guidelines provide a low microbe standard for unrestricted irrigation since it was based on epidemiological studies of developing countries where much of the population have now acquired immunity towards enteric infection [34,39]. On this note, the WHO [12] has since recommended new guidelines for wastewater irrigation which consider

Table 2. Irrigation water quality guidelines and standards for wastewater reuse in agriculture.

Parameters	South Korea [1] [57]		WHO [2] [12]		US EPA [33]		Cyprus [3] [26]		France [26]		Greece [35]		Israel [4] [36]	Italy [34]	Portugal [37]		Spain [38]	
Coliform (/100 mL)	Food Crops	ND TC	Unrestricted	*E. coli* (cfu) ≤ 1000	Food crops	ND FC (median)	Cooked vegetables	FC (MPN) ≤ 100	Unrestricted	*E. coli* (cfu) ≤ 250	Unrestricted	*E. coli* (cfu) ≤ 5 (80%) ≤ 50 (95%)	FC (cfu) ≤ 10	*E. coli* (cfu) ≤ 100 (max) ≤ 10 (80%)	Vegetables consumed raw [(a)]	FC (cfu) ≤ 100	Uncooked vegetables	*E. coli* (cfu) ≤ 100
	Processed food crops	TC (MPN) ≤ 200	Restricted	*E. coli* (cfu) ≤ 10,000	Processed food crops	FC (cfu) ≤ 200 (median)	Crops for human consumption	FC (MPN) ≤ 1000	All crops except those consumed raw	*E. coli* (cfu) ≤ 10,000	Restricted	*E. coli* (cfu) ≤ 200 (median)			Cooked vegetables	FC (cfu) ≤ 1000	Crops for human consumption	*E. coli* (cfu) ≤ 1000
Turbidity (NTU)	Food crops	≤2	- [(b)]		Food crops	≤2 (average)	-		-		Unrestricted	≤2 (median)	-	-	-		Uncooked vegetables	≤10
	Processed food crops	≤5			Processed food crops	-					Restricted	-		-			Crops for human consumption	-
Suspended solids (mg/L)	-		-		Food crops	-	Cooked vegetables	≤15	Unrestricted	<15	Unrestricted	≤10 (80%)	TSS ≤ 10	TSS ≤ 10	TSS ≤ 60		Uncooked vegetables	≤20
					Processed food crops	TSS ≤ 30	Crops for human consumption	≤45	All crops except those consumed raw	Varies [(c)]	Restricted	≤35					Crops for human consumption	≤35
BOD (mg/L)	≤8		-		Food crops	≤10	Cooked vegetables	≤15	-		Unrestricted	≤10 (80%)	≤10	≤20	-		-	-
					Processed food crops	≤30	Crops for human consumption	≤30			Restricted	≤25					-	-
COD (mg/L)	-		-		-		-		Unrestricted	<60	-		≤100	≤100	-		-	
									All crops except those consumed raw	Varies								
Odor	Do not unpleasant		-		-		-		-		-		-	-	-		-	
T-N (mg/L)	-		-		-		-		-		-		≤25	≤15	-		-	
T-P (mg/L)	-		-		-		-		-		-		≤5	≤2	-		-	
Intestinal nematodes (No./L)	-		≤1		-		ND		-		-		-	-	≤1		≤1(/10 L)	
pH	5.8–8.5		-		6.0–9.0		-		-		-		6.5–8.5	6.0–9.5	6.5–8.4		-	
EC (μs/cm)	Food crops	≤700	-		-		-		-		-		≤1400	≤3000	≤1000		-	
	Processed food crops	≤2,000																

ND = not detected; TC = total coliform; FC = fecal coliform; TSS = total suspended solids. [1] Standards for direct wastewater reuse. [2] The most stringent verification monitoring level, which refers to what has previously been referred to as effluent guideline levels, for each irrigation type and arithmetic mean value. [3] For vegetables eaten raw is not allowed and maximum value allowed. [4] Maximum monthly averages for unrestricted irrigation. [(a)] Vegetables whose edible parts are in close contact with the irrigated soil are not included and drip irrigation can be only employed. [(b)] No recommendation. [(c)] In accordance with wastewater treatment standards.

the human health risk through epidemiological studies and quantitative microbial risk assessment (QMRA), a process for estimating the risk of exposure to microorganisms (Table 3). The new recommended guidelines differ in microbe limit, depending on the irrigation method and crop type. It also takes into account the risk mitigation effect that comes from the entire process of agricultural production-from the irrigation system to the pre-consumption cleansing-in determining the permitted microbe limits.

Table 3. Recommended minimum verification monitoring of microbial performance targets for wastewater use in agriculture (modified from WHO [12]).

Type of Irrigation	*E. coli* (cfu/100 mL) (Arithmetic Mean)	Helminth Eggs (No./L) (Arithmetic Mean)
Unrestricted [1]		
Root crops [(a)]	$\leqslant 10^3$	
Leaf crops [(b)]	$\leqslant 10^4$	$\leqslant 1$
Drip irrigation, low-growing crops	$\leqslant 10^3$	
Drip irrigation, high-growing crops [(c)]	$\leqslant 10^5$	- [(d)]
Restricted [2]		
Labor-intensive, high-contact agriculture	$\leqslant 10^4$	$\leqslant 1$
Highly mechanized agriculture	$\leqslant 10^5$	$\leqslant 1$
Pathogen removal in a septic tank	$\leqslant 10^6$	$\leqslant 1$

[1] Use of treated wastewater to grow crops that are normally eaten raw. [2] Use of treated wastewater to grow crops that are not eaten raw by human. [(a)] Crops that may be eaten uncooked. [(b)] Vegetables eaten uncooked such as lettuce and cabbage. [(c)] Crops such as fruit trees and olives. [(d)] No recommendation.

3.2. US EPA

The US EPA adopts strict standards where it totally eliminates the risk of infection [39]. These strict standards require an excessive cost for the prevention of infectious diseases. Shuval *et al.* [58] estimated that the guidelines of US EPA, compared to those of WHO, will cost an additional $3 to $30 million per prevented enteric disease. In addition, there have been criticisms towards the strict US EPA guidelines since it might be impossible for developing countries to adopt them due to the required cost and technology [12].

3.3. Cyprus

Cyprus proclaimed water quality standards for wastewater reuse in 2005 [34] and is prohibiting the irrigation of treated wastewater for vegetables that are consumed raw, crops for exporting, and ornamental plants [24]. Water quality standards for vegetables that are consumed cooked and other crops are stricter than the WHO guidelines, but relaxed compared to the US EPA guidelines. Different water quality standards are recommended depending on the method of water treatment. The maximum allowable value and the value of which 80% of samples should not exceed are each presented.

3.4. France

In 1991, France issued wastewater reuse standards, which essentially follow the WHO guidelines [29] but add restrictions for irrigation techniques and setback distances to mitigate health risks [59]. Updated standards were devised in 2010, and they follow the principles of the revised WHO's guidelines [12] but include five other criteria for total suspended solids (TSS), chemical oxygen demand (COD), Enterococci, F-specific RNA phages, and spores of anaerobic bacteria in addition to limits for *E. coli* [26].

3.5. Greece

The water reuse standards of Greece took effect in 2011 and were advised to apply different standards for restricted irrigation and unrestricted irrigation in relation to subjects such as crop type, irrigation method, and accessibility of the public. Restricted irrigation refers to the method whereby sprinklers are not permitted and the irrigation area is inaccessible. This type of irrigation is subjected to crops that are either non-edible, consumed after being processed, or the yielding part is not in direct contact with the soil [35]. Unrestricted irrigation is independent from the irrigation method, applied to all types of crop, and also does not restrict public access to the irrigation area [35]. For certain water quality factors which require a strict standard, a probabilistic consideration is adopted for all water quality samples.

3.6. Israel

In 1952, Israel established legalized water quality standards for reusing wastewater for agriculture [54]. After that, new irrigation water quality standards for wastewater reuse-which require a more advanced wastewater treatment system than the existing-were devised in 2010 to minimize the damage of the water environment as well as crop and soil [60]. The new standards, to which factors such as the five-day BOD, TSS,

COD, fecal coliform, residual chlorine, and sodium adsorption ratio (SAR) are subjected, aim to use the entire amount of wastewater for unrestricted irrigation [36]. A notable change compared to the 1999 standards is the consideration of nutrients such as T-N, total phosphorus (T-P), residual chlorine, and trace elements such as boron (B).

3.7. Italy

Italy started to regulate the irrigation of treated wastewater by law in 1977, and new national standards were enacted in 2006 [25,59]. Microbiological factors such as enteric nematodes, which are normally considered important, were excluded in the new standards, whereas germs such as salmonella and trace elements were included. 20% of all the categories apply the same standard as drinking water, and 37% are categories which do not exist for drinking water quality standards [59]. The strict water quality standards of Italy are not differentiated by crop types or irrigation methods but can be a factor in limiting the broad use of wastewater reuse.

3.8. Portugal

The first standards on the use of treated wastewater for irrigation were published in Portugal in 2006 [37]. The standards distinguished into four classes according to the level of risk of microbiological contamination generated by wastewater irrigation [37]. The standards also provide guidelines for the selection of irrigation equipment and methods and for environmental protection and environmental impact monitoring in areas irrigated with treated wastewater [37]. According to the Portuguese standards, wastewater irrigation is not allowed in areas with high slopes (>20%) or very permeable soil (e.g., karstic areas), while the groundwater table should be at depth greater than 1–4 m during irrigation, depending on the type of irrigation method [37].

3.9. Spain

The water quality standards for wastewater reuse in Spain were established and proclaimed in 2007. Even before the establishment of nation-wide standards, the local governments of Catalonia, Valencia, and the Balearic Islands have advised their own standards based on the WHO water quality guidelines [38]. The recently established standards regulate the level of microbes and suspended solids: The regulation of germs such as Legionella is an especially notable feature.

4. WATER QUALITY STANDARDS FOR INDIRECT WASTEWATER REUSE

4.1. Comparison and Importance of Each Water Quality Criteria

4.1.1. Coliform Bacteria

Based on the total number of *E. coli*, the standards of California recommend a 7-day median value of 2.2 cfu/100 mL, and a maximum value of 240 cfu/100 mL. On the contrary, the WHO [12] guidelines recommend 1000 cfu/100 mL, even for the strictest standards for unrestricted irrigation. Since the number of *E. coli* is smaller than the number of total coliform, the relative strictness of the standards of California compared to that of WHO could be even greater. The water quality standards of most countries that are examined in this study adopt the concept of the standards of California, except for France, whose standards are based on the revised WHO guidelines. Portugal and Spain are also currently using standards that have been tightened compared to the WHO guidelines and can be classified as an intermediate-level regulation [26].

The current coliform bacteria regulation for direct wastewater reuse by the MOE is much stricter than that of the WHO. Compared to the US EPA, which only involves fecal coliform, the MOE, which involves total coliform, applies stricter water quality standards. Out of the 44 WWTPs which directly reuse the treated wastewater for agriculture in South Korea, 8 WWTPs exceed the coliform bacteria standard for processed food crops of 200 MPN/100 mL. The average total coliform for all 44 WWTPs was 139 MPN/100 mL [61]. Therefore, it is inappropriate to use treated wastewater to irrigate food crops with the current level of treatment and water quality standards. Even for processed food crops, additional treatment is needed. Especially, considering the current water quality standards that use the maximum allowable value and the treated wastewater quality report of the MOE, which is based on the annual average value, it can be said that most of the reused wastewater for agriculture exceeds the coliform bacteria standard.

Through examining the water quality of sites that adopt indirect wastewater reuse for paddy irrigation in South Korea, an average of 19,000 MPN/100 mL of total coliform was detected during the irrigation period [20]. Even for sites irrigated by relatively clean water derived from a reservoir, an average of about 15,700 MPN/100 mL of total coliform was detected [20]. It was also found that the total coliform of the irrigation water containing untreated wastewater was about 77,000 MPN/100 mL [21]. QMRA on paddy irrigation showed that the human health risk of irrigating treated wastewater is nine times higher than that of irrigating groundwater [42], but even with the extremely high level of total coliform, there have rarely been any reports of human harm from using treated or untreated wastewater for rice paddy irrigation. This indicates that the subjected level of total coliform can be easily found in nature and is inadequate as an indicator for microbes. The *E. coli* concentration of irrigation water with high levels of total coliform concentration was measured to be 13, 46, and 107 MPN/100 mL for reservoir water, indirect wastewater reuse water, and irrigation water containing untreated wastewater, respectively [20,21]. For fecal coliform, the concentrations for the subjects above were 5000, 180, and 633 MPN/100 mL, respectively [20,21]. Considering the high measurement of fecal coliform, even for conventional irrigation water derived from a reservoir, there needs to be a practical regulation using *E. coli*, which can well reflect the pathogenic microorganism such as the regulations of Spain, Greece, and Italy.

There have been numerous studies of human health risk regarding crops that are consumed raw, but, as has been proven through many QMRAs, results can vary depending on the initial conditions or the design of the experiment [62,63]. Therefore, there needs to be an attempt to lower the public concern of wastewater reuse practices and to eliminate human health risk by looking at the problem conservatively. To achieve such goals, it is recommended that reference countries adopt strict and conservative wastewater reuse standards, especially for crops that are consumed raw.

Generally, the paddy rice has a low risk of pathogenic bacteria infection since it goes through polishing processes and is consumed after being cooked. However, due to the nature of rice paddy cultivation, the irrigated water is contained for a certain period in paddy fields; therefore, when using treated wastewater for irrigation, issues of hygiene and safety of the farmer should be considered. Through QMRA of the *E. coli* concentration of paddy rice irrigation water, it was examined that the human health risk is very low when irrigating reclaimed wastewater. This means that additionally treated effluent after the regular WWTP processes meets the requirement for irrigation water [17]. However, when irrigating with treated wastewater diluted with stream water, the *E. coli* concentration exceeded the enteric disease risk value of 10^{-4}, and this has been reported to cause infection to farmers [17]. Yoon *et al.* [64] reported that the above risk can be drastically mitigated when farming activities take place 24 h after irrigation.

Therefore, an additional treatment process or safer farming practices for indirect wastewater reuse is necessary.

4.1.2. Salinity

Treated wastewater contains high salt content which can immensely affect crop growth; thus, there needs to be a standard for salinity [65]. Salinity is usually described in electric conductivity (EC). The effect that salinity gives to crop growth differs by crop type. Generally, if the EC of irrigation water is below 700 μs/cm, it does not affect crop growth; when above 3000 μs/cm, it can cause severe damage [45]. Israel and Italy aim for unrestricted irrigation and have EC standards for wastewater reuse of 1400 and 3000 μs/cm, respectively. Portugal has a fairly strict EC standard of 1000 μs/cm. In South Korea, the MOE sets the EC standard for direct wastewater reuse differently for food crops and processed food crops; the standards are 700 and 2000 μs/cm, respectively. When monitoring the average EC values of irrigation waters, the reservoir irrigation water displayed 170 μs/cm, indirect wastewater irrigation water 960 μs/cm, and irrigation water containing untreated wastewater 330 μs/cm [20,21]. The salinity level of indirect wastewater irrigation could exceed the directly consumed crop's tolerance level, and additional treatments are therefore needed. The allowed EC for paddy rice was reported to be 2000 μs/cm [66,67]. The EC standard of direct wastewater reuse for processed food crops can thus be applied to that of indirect reuse for rice paddy irrigation.

4.1.3. Turbidity or Suspended Solids

Other than South Korea, countries that advise a standard for turbidity are the US EPA, Spain, and Greece, where the standards are usually subjected to unrestricted irrigation. The US EPA and Greece recommend a level lower than 2 NTU only for directly consumed crops and unrestricted irrigation, and Spain recommends a level lower than 10 NTU for vegetables. The MOE sets the standards at under 2 NTU for food crops, and under 5 NTU for processed food crops. Most countries that do not have a standard for turbidity have a standard for suspended solids instead. Spain and Greece have both standards for turbidity and suspended solids regarding irrigation water for vegetables.

A high level of turbidity can affect the performance of the irrigation facility, and can lower the hydraulic conductivity of the soil and in turn pollute the soil surface through surface flow [68,69]. In addition, since various viruses and bacteria can be attached to and migrate along with the solid particles, the elimination of suspended solids is related to the elimination of germs [33]. Therefore, the standard for turbidity can be set up based on the turbidity's influence on the irrigation facility performance, or vegetables which are vulnerable to germ infection. In the case of indirect wastewater reuse, a strict standard of 2 NTU can be applied for directly consumed crops, and, for indirectly consumed crops, a specific standard that can prevent the adverse effects is needed. Many countries apply the suspended solids standards for indirectly consumed crops, and South Korea can use the 15 mg/L standard, which is a standard for using lake water as agricultural irrigation. In the case of paddy rice, where water is supplied through surface irrigation, adverse effects to irrigation facilities do not exist, and the human health risk can be controlled through a microbe standard. Thus, it can be said that an additional standard regarding suspended solids is not necessary.

4.1.4. Organic Matter

Organic matters not only negatively affect the odor and color of the water, but also act as nutrients for microbes and bring various adverse effects during the disinfection process [33]. Israel and Italy, which have the most strict water quality standards for wastewater reuse, have standards for BOD and COD. However, there are cases such as the US EPA and France where they only have regulations for BOD and COD, respectively, while there are no standards for organic matters at all in WHO, Portugal, or Spain. Even when comparing with Israel's 10 mg/L, South Korea has a very strict BOD standard because of the emphasis on the regulation in terms of stream water quality management. In examining water quality of the treated wastewater effluents from the 44 WWTPs of South Korea, the average BOD measurement was 2.6 mg/L [61]. The BOD measurement of irrigation water as indirect wastewater reuse was 2.0 mg/L [20]; even for irrigation water containing untreated wastewater, BOD showed an extremely low level of 2.6 mg/L [21]. Therefore, it can be said that applying the current strict standards of the MOE for direct wastewater reuse is adequate enough even considering the regional idiosyncrasies.

4.1.5. Nutrients

Israel and Italy have set standards for nutrients regarding wastewater reuse. Israel sets standards for T-N and T-P as 25 and 5 mg/L, respectively, and the same standards for Italy are 15 and 2 mg/L, respectively. South Korea does not have standards for nutrients specifically for wastewater reuse, but nevertheless have strict water quality standards for stream water quality management: The regulation levels of T-N and T-P are 1 and 0.1 mg/L, respectively. When examining the reservoir irrigation water in South Korea, results confirmed a T-N value of 1.1 mg/L and a T-P value of 0.05 mg/L, which are relatively good [21]. In the case of indirect wastewater reuse, T-N and T-P were measured at 3.3 and 1.6 mg/L, respectively [20]. Irrigation water containing untreated wastewater confirmed T-N and T-P values of 14.0 and 2.8 mg/L, respectively [21]. The above results, even when examined against the relatively strict standards of Israel and Italy, are fairly descent.

In South Korea, during the dry season of May to June, which is when most of the irrigation activity occurs, irrigation water by indirect wastewater reuse contains a high level of T-N concentration due to the decrease in streamflow volume [14]. This can lead to a decrease in production due to lodging [46,70]. In consideration of the above negative effects of nitrogen on paddy rice growth, Taiwan and Japan have set standards for T-N to be under 3.0 [46] and 1.0 mg/L [70], respectively. Therefore, for better quality of paddy rice, a stricter standard should be applied. Depending on circumstances, a lowering of T-N concentration via additional treatment or via an application of appropriate fertilization practices in paddy farming might also be needed [10]. Phosphate can stimulate the growth of bacteria or algae, which can cause the clogging of irrigation facilities [71]; however, since the South Korean standard for T-P of effluent discharge is strict (being 0.2–2.0 mg/L), and considering the fact that effluent discharge from a WWTP in South Korea is usually used after being diluted with lake or stream water, there seems to be no need for additional restrictions for phosphate.

4.1.6. pH

The US EPA recommends the appropriate pH range to be 6.0–9.0, Israel 6.5–9.5, Italy 6.0–9.5, and Portugal 6.5–8.4. The MOE sets the pH standard for direct wastewater reuse as 5.8–8.5, the lower limit being somewhat low compared to the standards of other countries. Low pH values affect the mobility of heavy metals in the soil and can be absorbed by crops and contaminate water bodies [12]. Taking into account the existing agricultural water standards of pH 6.0–8.5 applied to stream, lake, and groundwater in South Korea, and considering the effect of pH on irrigation facilities, raising the lower limit of the current pH standard for the direct wastewater reuse seems appropriate for upland crops, which requires additional irrigation facilities.

4.1.7. Trace Elements

While the South Korean standards for direct wastewater reuse regarding trace elements are stricter than that of FAO [45] or US EPA [33] and similar to those of Cyprus, Greece, Israel, and Italy, it lacks the standards for elements such as beryllium (Be), fluorine (F), iron (Fe), molybdenum (Mo), and vanadium (V), for which many countries have standards (Table 4). Be, although having different effects on crops depending on the crop type, can be consumed by humans as a carcinogen through root crop, tuber crop, and forage crop [72]. F can damage the crops by accumulating in the leaf and fruit [73]. Fe, while not harmful for crops in aerated soil, can cause immense damage by decreasing the vital P in irrigated paddy of acid soil [74,75]. In addition, the forage crops which grow in soil with a high concentration of Mo can be poisonous to livestock [76], and V is known to be harmful to many crops even in relatively low concentrations [33].

Table 4. Comparison of recommended maximum concentrations of trace elements in irrigation water.

Parameters	Korea [56]	FAO [45]	US EPA [33]	Cyprus [25]	Greece [25]	Israel [36]	Italy [34]
Aluminum, Al	5.0	5.0	5.0	5.0	5.0	5.0	1.0
Arsenic, As	0.05	0.1	0.1	0.1	0.1	0.1	0.02
Beryllium, Be	- [(a)]	0.1	0.1	0.1	0.1	0.1	-
Boron, B	0.75	0.7	0.75	0.75	2.0	0.4	1.0
Cadmium, Cd	0.01	0.01	0.01	0.01	0.01	0.01	0.005
Chromium, Cr	0.05	0.1	0.1	0.1	0.1	0.1	0.1
Cobalt, Co	0.05	0.05	0.05	0.05	0.05	0.05	0.05
Copper, Cu	0.2	0.2	0.2	0.2	0.2	0.2	1.0
Cyanide, CN	ND [(b)]	–	–	–	–	0.1	0.05
Fluoride, F	–	1.0	1.0	–	1.0	2.0	1.5
Iron, Fe	–	5.0	5.0	5.0	3.0	2.0	2.0
Lead, Pb	0.1	5.0	5.0	5.0	0.1	0.1	0.1
Lithium, Li	2.5	2.5	2.5	2.5	2.5	2.5	–
Manganese, Mn	0.2	0.2	0.2	0.2	0.2	0.2	0.2
Mercury, Hg	0.001	–	–	–	0.002	0.002	0.001
Molybdenum, Mo	–	0.01	0.01	0.01	0.01	0.01	–
Nickel, Ni	0.2	0.2	0.2	0.2	0.2	0.2	0.2
Selenium, Se	0.02	0.02	0.02	0.02	0.02	0.02	0.01
Tin, Sn	–	–	–	–	–	–	3.0
Thallium, Tl	–	–	–	–	–	–	0.001
Vanadium, V	–	0.1	0.2	2.0	0.1	0.1	0.1
Zinc, Zn	2.0	2.0	2.0	0.005	2.0	2.0	0.5

[(a)] No recommendation; [(b)] Not detected.

Treated wastewater after secondary treatment is adequate for reuse since the level of heavy metals in the effluent is similar to that in nature [45]. In monitoring the water quality of three types of irrigation water in regards to categories of Al, Cu, Ni, Zn, As, Cd, Cr, Co, Pb, Li, Mn, Hg, Se, and CN, reservoir water did not exceed any of the direct wastewater reuse standards, indirectly reused wastewater exceeded levels of Al and Mn, and irrigation water containing untreated wastewater exceeded the levels of Al, Cu, Zn, and Mn [20,21]. Considering the average concentration during the irrigation period and the standards of other countries, the detected high levels of Cu and Zn were negligible, whereas the results for Al and Mn were concerning. Additional treatment for Al and Mn is needed since they can be poisonous to crops or decrease crop yields [52]. South Korea should set standards for elements such as Be, F,

Fe, Mo, and V considering the specific conditions of the country; since there are no preceding studies on the subject, it seems practical to adopt the standards of other countries.

4.1.8. Odor

There are normally no specific standards for odor in agricultural wastewater reuse, but the South Korean standards state that the odor "should not be unpleasant", and this could have a positive effect of reducing the animosity towards the idea of wastewater reuse. The perception of farmers towards wastewater reuse can be an important criterion in establishing water quality standards [77], and physical characteristics such as odor have a strong influence on farmers' perceptions. Therefore, including a factor of odor for water quality standards could be a good approach in considering stakeholders.

4.1.9. Level of Treatment

The US EPA [33] sets the water quality guidelines so that wastewater is disinfected to a certain level regardless of the purpose of use. This is to prevent side effects to the human body that might occur from accidental contact with the treated wastewater. However, a water treatment technology to fulfill the given water quality standards is inevitable, albeit the absence of specific treatment guidelines for wastewater reuse. Israel and Italy require advanced and expensive treatment systems to meet its water quality standards [34,36], and Portugal also has treatment requirements as per its standards [37]. South Korea also requires more than secondary treatment for direct wastewater reuse, and, depending on the occasion, RO (Reverse Osmosis) and advanced treatment are applied as well.

4.2. Draft Water Quality Standards for Indirect Wastewater Reuse in South Korea

4.2.1. Differentiation by Crop Type

Generally, the water quality standards for wastewater irrigation are established depending on the type of subjected crop. The WHO's differentiation of unrestricted irrigation and restricted irrigation is based on whether the subjected crop is consumed raw or not, and the US EPA [33] also classifies crops into either food crops or non-food/processed food crops. Cyprus sets different water quality standards for processed food vegetables, human consumed crops, forage crops, and industrial crops. Portugal only has different standards of a microbiological parameter (FC) according to a crop's intended use. Spain has separate standards for raw consumed vegetables, human consumed crops, and industrial crops. Israel and Italy, due to their goal of unrestricted irrigation, do not differentiate water quality standards based on crop type. France sets different standards for four effluent qualities based on its potential use for crops and the accessibility of the public. Greece applies different standards of restricted irrigation and unrestricted irrigation based on not only crop type, but also the irrigation method and the accessibility of the public. The MOE sets different standards for direct wastewater reuse based on whether the subjected crop can be consumed raw or not.

Indirect wastewater reuse is usually practiced haphazardly, and this indicates that the water quality cannot be manually controlled unless the treated wastewater is additionally treated. Therefore, introducing an economical treatment system seems implausible in the case of rice paddy, where a large quantity of irrigation water is required. Still, in the case of upland crops, including vegetables raised through protected cultivation, the irrigation water quality standards need to consider the way the crop is consumed and the irrigation method. In addition, considering the change in the farming environment and the economic feasibility of using agricultural water, specific standards should be established for fodder crops and industrial crops as well.

4.2.2. Probabilistic Consideration

Like France, Portugal, and Spain, the water quality standards for direct wastewater reuse of South Korea use the maximum allowable value, yet are much stricter than those countries. The WHO [12] uses arithmetic mean as criteria for the water quality guidelines, and Israel uses the monthly average. The water quality guidelines of the US EPA use the median or average value depending on the subjected category. The state standards of the US are based on either geometric mean or exceeded number per sample, and, in the case of fecal coliform, apply additional factors of median value or a greatly mitigated maximum allowable value. Greece and Italy apply probabilistic considerations to their water quality standards. Each monitored value of the water quality categories are greatly influenced by the hydrological changes of the environment at which the samples are collected; therefore, using a strict maximum allowable value does not well reflect reality. Thus, as most countries do, applying the average value or the exceedance probability to the water quality standards is recommended.

4.2.3. Draft Water Quality Standards

From the above examination, a draft proposal for the water quality standards of indirect wastewater reuse was suggested, as seen in Table 5. The proposed water quality standards, while referencing the existing standards of direct wastewater reuse as well as reflecting the farming environment of South Korea, classified the use of irrigation water into rice paddy and upland. The standards also considered the categories of *E.*

coli, EC, turbidity, SS, BOD, pH, and odor. Additionally, for trace elements, categories of the FAO standards such as Be, F, Fe, Mo, and V were added to the existing direct wastewater reuse standards. Wastewater treatment system or reuse system is needed to satisfy the water quality standards if irrigation water quality exceeds the standards.

Table 5. A draft proposal for the water quality standards of indirect wastewater reuse in South Korea.

Parameters	Upland Irrigation		Rice Paddy Irrigation [(a)]
	Food Crops	Processed Food Crops	
E. coli (MPN/100 mL)	⩽10 [(b)]	⩽10 (monthly mean) ⩽ 200 (max)	⩽1000 (max)
EC (μs/cm)	⩽700 (max)	⩽2000 (max)	⩽2000 (max)
Turbidity (NTU)	⩽2 (monthly mean)	– [(c)]	–
Suspended solids (mg/L)	–	⩽15 (monthly mean)	–
BOD (mg/L)	⩽8 (monthly mean)		
pH	6.0–8.5		
Odor	Do not unpleasant		
Al, As, B, Cd, Cr, Co, Cu, CN, Pb, Li, Mn, Hg, Ni, Se, Zn	Korean standards for direct wastewater reuse standards [(d)]		
Be, F, Fe, Mo, V	FAO standards [(e)]		

[(a)] Agricultural activities should be carried out 24 h later after wastewater irrigation and 20% to 50% reduced fertilizer rate compared to the standard fertilizer rate is recommended. [(b)] 80% of samples below detection. [(c)] No recommendation. [(d)] The Korean standards for direct wastewater reuse standards were presented in Table 4. [(e)] The FAO standards were presented in Table 4.

The conservative Californian approach was adopted for the irrigation water of upland crops which can be consumed raw. For rice paddy irrigation water, the liberal approach of the WHO was adopted to prepare for the changes in water resources-such as water shortage and climate change-by vitalizing wastewater reuse for rice paddy irrigation, which takes up most of the freshwater resource demand. This kind of approach corresponds to the endeavor of Angelakis *et al.* [23], who asserted a third way which hybridizes a conservative approach and a liberal approach.

4.3. Implication of the Proposed Standards

There can be many kinds of agricultural wastewater reuse, depending on the different agricultural environments and types of agricultural practices. Due to the lack of information on wastewater reuse environments, especially in developing countries, it is hard to devise an appropriate policy which can promote the practice of sustainable wastewater reuse. Indirect wastewater reuse-which is practiced more commonly in water-rich environments-and rice paddy-oriented agriculture are the key characteristics of wastewater reuse in South and

East Asia. These regions, unfortunately, face serious water quality issues, which in turn cause freshwater scarcity, ill-health, and even fatalities [78]. Agricultural water is gradually deteriorating mainly due to polluted domestic and industrial effluents. Many of the rivers in South and East Asia contain up to three times the world average of fecal bacteria measure [79]. Asian countries are making concerted efforts to address these problems [78], and it was subsequently addressed by a variety of measures, including statutory regulation of water pollution control and other legislations [80]. Technologies and guidelines for indirect wastewater reuse could be appropriate measures for the safe and sustainable practices regarding water resources. In addition, international political pledges have been made at the national and local levels [80]. Therefore, the proposed draft standards would contribute to overcoming water problems by providing suitable criteria for practicing sustainable wastewater reuse in Asia.

5. CONCLUSIONS

Indirect wastewater reuse has been increasing due to urbanization and the increase of WWTPs, which in turn influences the irrigation water of the treated wastewater. For safe and sustainable practices of wastewater reuse, there needs to be appropriate water quality standards for indirect wastewater reuse. In this study, draft standards were proposed referencing the standards of the countries which already have, or have modified, water quality standards for agricultural wastewater reuse. In doing so, the water quality requirements for irrigation water and wastewater reuse were examined, and the standards of other countries on wastewater reuse were compared and analyzed. The proposed standards for indirect wastewater reuse classified the use of irrigation water into rice paddy and upland, and adopt probabilistic considerations for their practicality of use. Categories of *E. coli*, EC, turbidity, SS, BOD, pH, and odor were also considered, and trace elements such as Be, F, Fe, Mo, and V, which did not exist in the current standards for direct wastewater reuse, were added. The conservative Californian approach was adopted for the irrigation water of upland crops, and the liberal approach of the WHO was adopted for paddy irrigation water: Thus, a hybridization of the two was attempted. The proposed standards of this study are expected to vitalize wastewater reuse for rice paddy irrigation, which constitutes most of the freshwater resource demand, and to provide suitable criteria for practicing sustainable wastewater reuse.

ACKNOWLEDGMENTS

This work was supported by the Korea Institute of Planning and Evaluation for Technology in Food, Agriculture, Forestry and Fisheries (IPET) through Agri-Bio industry Technology Development Program, funded by Ministry of Agriculture, Food, and Rural Affairs (MAFRA) (grant number 114060-3).

AUTHOR CONTRIBUTIONS

Hanseok Jeong and Hakkwan Kim developed the conceptual framework for this manuscript, collected and analyzed the data for the standards and guidelines of various countries and organizations, suggested the appropriate water quality standards, produced the tables, and wrote the paper. Taeil Jang helped to choose a proper method for data display, suggested the appropriate water quality standards, and produced the final tables. Hanseok Jeong, Hakkwan Kim, and Taeil Jang all read and made improvements to the manuscript.

CONFLICTS OF INTEREST

The authors declare no conflicts of interest.

REFERENCES

1. Winpenny, J.; Heinz, I.; Koo-Oshima, S. *The Wealth of Waste: The Economics of Wastewater Use in Agriculture*; Water Reports; Food and Agriculture Organization: Rome, Italy, 2010.
2. Hoekstra, A.Y.; Mekonnen, M.M. The water footprint of humanity. *Proc. Natl. Acad. Sci. USA* **2012**,*109*, 3232–3237.
3. Hong, E.M.; Choi, J.Y.; Nam, W.H.; Kang, M.S.; Jang, J.R. Monitoring nutrient accumulation and leaching in plastic greenhouse cultivation. *Agric. Water Manag.* **2014**, *146*, 11–23.
4. Nam, W.H.; Choi, J.Y.; Hong, E.M. Irrigation vulnerability assessment on agricultural water supply risk for adaptive management of climate change in South Korea. *Agric. Water Manag.* **2015**,*152*, 173–187.

5. Bixio, D.; Thoeye, C.; De Koning, J.; Joksimovic, D.; Savic, D.; Wintgens, T.; Melin, T. Wastewater reuse in Europe. *Desalination* **2006**, *187*, 89–101.

6. Huertas, E.; Salgot, M.; Hollender, J.; Weber, S.; Dott, W.; Khan, S.; Schafer, A.; Messalem, R.; Bis, B.; Aharoni, A.; Chikurel, H. Key objectives for water reuse concepts. *Desalination* **2008**, *218*, 120–131.

7. Pedrero, F.; Kalavrouziotis, I.; Alarcon, J.J.; Koukoulakis, P.; Asano, T. Use of treated municipal wastewater in irrigated agriculture-review of some practices in Spain and Greece. *Agric. Water Manag.* **2010**, *97*, 1233–1241.

8. Chavez, A.; Rodas, K.; Prado, B.; Thompson, R.; Jimenez, B. An evaluation of the effects of changing wastewater irrigation regime for the production of alfalfa (Medicago sativa). *Agric. Water Manag.* **2012**, *113*, 76–84.

9. Jang, T.I.; Kim, H.K.; Seong, C.H.; Lee, E.J.; Park, S.W. Assessing nutrient losses of reclaimed wastewater irrigation in paddy fields for sustainable agriculture. *Agric. Water Manag.* **2012**, *104*, 235–243.

10. Jeong, H.S.; Jang, T.I.; Seong, C.H.; Park, S.W. Assessing nitrogen fertilizer rates and split applications using the DSSAT model for rice irrigated with urban wastewater. *Agric. Water Manag.***2014**, *141*, 1–9.

11. Hamilton, A.J.; Stagnitti, F.; Xiong, X.; Kreidil, S.L.; Benke, K.K.; Maher, P. Wastewater irrigation: The state of play. *Vadose Zone J.* **2007**, *6*, 823–840.

12. *Guidelines for the Safe Use of Wastewater, Excreta and Greywater*; World Health Organization: Geneva, Switzerland, 2006.

13. Rutkowski, T.; Raschid-Sally, L.; Buechler, S. Wastewater irrigation in the developing world-Two case studies from the Kathmandu Valley in Nepal. *Agric. Water Manag.* **2007**, *88*, 83–91.

14. Jeong, H.; Kim, H.; Jang, T.; Park, S. Assessing the effects of indirect wastewater reuse on paddy irrigation in the Osan River watershed in Korea using the SWAT model. *Agric. Water Manag.* **2016**,*163*, 393–402.

15. Mohammad, M.J.; Mazahreh, N. Changes in soil fertility parameters in response to irrigation of forage crops with

secondary treated wastewater. *Commun. Soil Sci. Plant Anal.* **2003**, *34*, 1281–1294.

16. Xu, J.; Wu, L.; Chang, A.C.; Zhang, Y. Impact of long term reclaimed wastewater irrigation on agricultural soils: A preliminary assessment. *J. Hazard. Mater.* **2010**, *183*, 780–786.

17. Rhee, H.P.; Yoon, C.G.; Jung, K.W.; Son, J.W. Microbial risk assessment using *E. coli* in UV disinfected wastewater irrigation on paddy. *Environ. Eng. Res.* **2009**, *14*, 120–125.

18. Forslund, A.; Ensink, J.H.J.; Battilani, A.; Kljujev, I.; Gola, S.; Raicevic, V.; Jovanovic, Z.; Stikic, R.; Sandei, L.; Fletcher, T.; *et al.* Faecal contamination and hygiene aspect associated with the use of treated wastewater and canal water for irrigation of potatoes. *Agric. Water Manag.* **2010**, *98*, 440–450.

19. Banon, S.; Miralles, J.; Ochoa, J.; Franco, J.A.; Sanchez-Blanco, M.J. Effects of diluted and undiluted treated wastewater on the growth, physiological aspects and visual quality of potted lantana and polygala plants. *Sci. Hortic.* **2011**, *129*, 869–876.

20. Jeong, H.S.; Park, J.H.; Seong, C.H.; Jang, T.I.; Kang, M.S.; Park, S.W. Effects of indirect wastewater reuse on water quality and soil environment in paddy fields. *J. Korean Soc. Agric. Eng.* **2013**, *55*, 91–104. (In Korean)

21. Song, J.H.; Jeong, H.S.; Park, J.H.; Song, I.H.; Kang, M.S.; Park, S.W. Analysis of water quality and soil environment in paddy fields partially irrigated with untreated wastewater. *J. Korean Soc. Agric. Eng.* **2014**, *56*, 19–29. (In Korean)

22. Kim, J.H.; Jeong, H.S.; Kang, M.S.; Song, I.H.; Park, S.W. Simulation of 10-day irrigation water quality using SWAT-QUALKO2 linkage model. *J. Korean Soc. Agric. Eng.* **2012**, *54*, 53–63. (In Korean)

23. Angelakis, A.N.; Marecos do Monte, M.H.F.; Bontoux, L.; Asano, T. The status of wastewater reuse practice in the Mediterranean basin: Need for guidelines. *Water Res.* **1999**, *33*, 2201–2217.

24. Brissaud, F. Criteria for water recycling and reuse in the Mediterranean countries. *Desalination***2008**, *218*, 24–33.

25. Kalavrouziotis, I.K.; Kokkinos, P.; Oron, G.; Fatone, F.; Bolzonella, D.; Vatyliotou, M.; Fatta-Kassinos, D.; Koukoulakis, P.H.; Varnavas, S.P. Current status in wastewater treatment, reuse and research in some mediterranean countries. *Desalin. Water Treat.* **2015**, *53*, 2015–2030.

26. Paranychianakis, N.V.; Salgot, M.; Snyder, S.A.; Angelakis, A.N. Water reuse in EU states: Necessity for uniform criteria to mitigate human and environmental risks. *Crit. Rev. Environ. Sci. Technol.* **2015**, *45*, 1409–1468.

27. Kretschmer, N.; Ribbe, L.; Gaese, H. Wastewater Reuse for Agriculture. *Technol. Resour. Manag. Dev. Sci. Contrib. Sustain. Dev.* **2002**, *2*, 37–64.

28. Jimenez, B. Treatment technology and standards for agricultural wastewater reuse: A case study in Mexico. *Irrig. Drain.* **2005**, *54*, S22–S33.

29. *Guidelines for the Safe Use of Waste Water and Excreta in Agriculture and Aquaculture: Measures for Public Health Protection*; World Health Organization: Geneva, Switzerland, 1989.

30. *Protocol Development: Criteria and Standards for Potable Reuse and Feasible Alternatives*; Environmental Protection Agency: Washington, DC, USA, 1980.

31. *Guidelines for Water Reuse 625/R92004*; Environmental Protection Agency: Washington, DC, USA, 1992.

32. *Guidelines for Water Reuse 625/R-04/108*; Environmental Protection Agency: Washington, DC, USA, 2004.

33. *Guidelines for Water Reuse 600/R-12/618*; Environmental Protection Agency: Washington, DC, USA, 2012.

34. Angelakis, A.N.; Durham, B.; Marecos Do Monte, M.H.F.; Salgot, M.; Witgens, T.; Thoeye, C. Wastewater recycling and reuse in Eureau countries: Trends and challenges. *Desalination* **2008**,*218*, 3–12.

35. Agrafioti, E.; Diamadopoulos, E. A strategic plan for reuse of treated municipal wastewater for crop irrigation on the Island of Crete. *Agric. Water Manag.* **2012**, *105*, 57–64.

36. *Sustainable Water Integrated Management-Support Mechanism (SWIM-SM)*; 4th Progress Report; LDK Consulatants Engineers and Planners SA: Athens, Greece, October 2013.

37. Marecos do Monte, M.H.F. Guidelines for good practice of water reuse for irrigation: Portuguese standard NP 4434. In *Wastewater Reuse-Risk Assessment, Decision-Making and Environmental Security*; Zaidi, M., Ed.; Springer: Dordrecht, The Netherlands, 2007; pp. 253–265.

38. Ortega, E.; Iglesias, R. Reuse of treated municipal wastewater effluents in Spain: Regulations and most common technologies, including extensive treatments. *Desalin. Water Treat.* **2009**, *4*, 148–160.

39. Blumenthal, U.J.; Peasy, A.; Ruiz-Palacios, G.; Mara, D.D. *Guidelines for Wastewater Reuse in Agriculture and Aquaculture: Recommended Revisions Based on New Research Evidence*; Well Resource Centre: London, UK, 2000.

40. Khepar, S.D.; Yadav, A.K.; Sondhi, S.K.; Siag, M. Water balance model for paddy fields under intermittent irrigation practices. *Irrig. Sci.* **2000**, *19*, 199–208.

41. Barker, R.; Dawe, D.; Tuong, T.P.; Bhuiyan, S.I.; Guerra, L.C. The outlook for water resources in the year 2020: Challenges for research on water management in rice production. *Southeast Asia***1999**, *1*, 1–5.

42. Jang, T.; Lee, S.B.; Sung, C.H.; Lee, H.P.; Park, S.W. Safe application of reclaimed water reuse for agriculture in Kora. *Paddy Water Environ.* **2010**, *8*, 227–233.

43. Beltran, J.M. Irrigation with saline water: Benefits and environmental impact. *Agric. Water Manag.***1999**, *40*, 183–194.

44. Bauder, T.A.; Waskom, R.M.; Sutherland, P.L.; Davis, J.G. *Irrigation Water Quality Criteria*; Colorado State University Extension: Fort Collins, CO, USA, 2011.

45. Ayers, R.S.; Westcot, D.W. *Water Quality for Agriculture*; Food and Agriculture Organization of the United Nations: Rome, Italy, 1985.

46. Setter, T.L.; Laureles, E.V.; Mazaredo, A.M. Lodging reduces yield of rice by self-shading and reductions in canopy photosynthesis. *Field Crop. Res.* **1997**, *49*, 95–106.

47. Chiou, R.J. Risk assessment and loading capacity of reclaimed waste-water to be reused for agricultural irrigation. *Environ. Monit. Assess.* **2008**, *142*, 255–262.

48. Kaneki, R. Reduction of effluent nitrogen and phosphorus from paddy fields. *Paddy Water Environ.* **2003**, *1*, 133–138.

49. Smith, V.H.; Tilman, G.D.; Nekola, J.C. Eutrophication: Impacts of excess nutrient inputs on freshwater, marine, and terrestrial ecosystems. *Environ. Pollut.* **1999**, *100*, 179–196.

50. Ju, X.T.; Kou, C.L.; Zhang, F.S.; Christie, P. Nitrogen balance and groundwater nitrate contamination: Comparison among three intensive cropping systems on the North China Plain.*Environ. Pollut.* **2006**, *143*, 117–125.

51. An, Y.J.; Lee, W.M.; Yoon, C.G. Evaluation of Korean water quality standards and suggestion of additional water parameters. *Korean J. Limnol.* **2006**, *39*, 285–295. (In Korean)

52. Asano, T.; Burton, F.L.; Leverenz, H.L.; Tsuchihashi, R.; Tchobanoglous, G. *Water Reuse: Issues, Technologies and Applications*; McGraw-Hill: New York, NY, USA, 2007.

53. Gupta, U.C.; Gupta, S.C. Trace element toxicity relationships to crop production and livestock and human health: Implications for management. *Commun. Soil Sci. Plant Anal.* **1998**, *29*, 1491–1522.

54. Salgot, M.; Angelakis, A.N. Guidelines and regulations on wastewater reuse. In *Decentralized Sanitation and Reuse: Concepts, Systems and Implementation*; Lens, P., Zeeman, G., Lettinga, G., Eds.; IWA Publishing: London, UK, 2001; pp. 446–466.

55. Tsagarakis, T.P.; Dialynas, G.E.; Angelakis, A.N. Water resources management in Crete (Greece) including water recycling and reuse and proposed quality criteria. *Agric. Water Manag.* **2004**, *66*, 35–47.

56. Nrmmc, E. *National Guidelines for Water Recycling: Managing Health and Environmental Risks*; the Environment Protection and Heritage Council and Natural Resource Management Ministerial Council: Canberra, Australia, 2006.

57. *Water Quality Standards for Treated Wastewater Based on Its Specific Purpose of Use*; Ministry of Environment: Gyeonggi-do, QKorea, 2011.

58. Shuval, H.; Lampert, Y.; Fattal, B. Development of a risk assessment approach for evaluating wastewater reuse standards for agriculture. *Water Sci. Technol.* **1997**, *35*, 15–20.

59. Raso, J. *Update of the Final Report on Wastewater Reuse in the European Union. Project: Service Contract for the Support to the Follow-up of the Communication on Water Scarcity and Droughts*; TYPSA Consulting Engineers and Architects: Barcelona, Spain, 2013.

60. Inbar, Y. New standards for treated wastewater reuse in Israel. In *Wastewater Reuse-Risk Assessment, Decision-Making and Environmental Security*; Zaidi, M., Ed.; Springer: Dordrecht, The Netherlands, 2007; pp. 291–296.

61. *Statistics of Sewerage*; Ministry of Environment: Sejong, Korea, 2014.

62. Petterson, S.R.; Ashbolt, N.; Sharma, A. Microbial risks from wastewater irrigation of salad crops: A screening-level risk assessment. *Water Environ. Res.* **2001**, *72*, 667–672.

63. Hamilton, A.J.; Stagnitti, F.; Premier, R.; Boland, A.M.; Hale, G. Quantitative microbial risk assessment models for consumption of raw vegetables irrigated with reclaimed water. *Appl. Environ. Microbiol.* **2006**, *72*, 3284–3290.

64. Yoon, C.G.; Han, J.Y.; Jung, K.W.; Jang, J.H. Quantitative microbial risk assessment of wastewater reuse for irrigation in paddy field. *J. Korean Soc. Agric. Eng.* **2006**, *48*, 77–87. (In Korean)

65. Kiziloglu, F.M.; Turan, M.; Sahin, U.; Kuslu, Y.; Dursun, A. Effects of untreated and treated wastewater irrigation on some chemical properties of cauliflower (Brassica olerecea L. var. botrytis) and red cabbage (Brassica olerecea L. var. rubra) grown on calcareous soil in Turkey.*Agric. Water Manag.* **2008**, *95*, 716–724.

66. Asch, F.; Wopereis, M.C.S. Responses of field-grown irrigated rice cultivars to varying levels of floodwater salinity in a semi-arid environment. *Field Crop. Res.* **2001**, *70*, 121–137.

67. Grattan, S.R.; Zeng, L.; Shannon, M.C.; Roberts, S.R. Rice is more sensitive to salinity than previously thought. *Calif. Agric.* **2002**, *56*, 189–195.

68. Vinten, A.J.A.; Minelgrin, U.; Yaron, B. The effect of suspended solids in wastewater on soil hydraulic conductivity: II. Vertical distribution of suspended solids. *Soil Sci. Soc. Am. J.* **1983**, *47*, 408–412.

69. Ragusa, S.R.; de Zoysa, D.S.; Rengasamy, P. The effect of microorganisms, salinity and turbidity on hydraulic conductivity of irrigation channel soil. *Irrig. Sci.* **1994**, *15*, 159–166.

70. Hidaka, S. Studies on effect of irrigation water quality upon the rice growth and soil, and its usable marginal concentration. *Bull. Saitama Agric. Exp. Station* **1990**, *44*, 1–100. (In Japanese)

71. Shatanawi, M.; Fayyad, M. Effect of Khirbet As-Samra treated effluent on the quality of irrigation water in the central Jordan Valley. *Water Res.* **1996**, *30*, 2915–2920.

72. Vaessen, H.A.; Szteke, B. Beryllium in food and drinking water: A summary of available knowledge. *Food Addit. Contam.* **2000**, *17*, 149–159.

73. Bustingorii, C.; Lavado, R.S. Soybean as affected by high concentrations of arsenic and fluoride in irrigation water in controlled conditions. *Agric. Water Manag.* **2014**, *144*, 134–139.

74. Fageria, N.K.; Baligar, V.C.; Clark, R.B. Micronutrients in crop production. *Adv. Agron.* **2002**, *77*, 185–268.

75. Fageria, N.K.; Baligar, V.C.; Wright, R.J. Iron nutrition of plants: An overview on the chemistry and physiology of its deficiency and toxicity. *Pesq. Agropec. Bras.* **1990**, *25*, 553–570.

76. Welch, R.M.; Allaway, W.H.; House, W.A.; Kubota, J. Geographic distribution of trace element problems. In *Micronutrients in Agriculture*, 2nd ed.; Mortvedt, J.J., Cox, F.R., Shuman, L.M., Welch, R.M., Eds.; Soil Science Society America: Madison, WI, USA, 1991; pp. 31–57.

77. Friedler, E.; Lahav, O.; Jizhaki, H.; Lahav, T. Study of urban population attitudes towards various wastewater reuse options: Israel as a case study. *J. Environ. Manag.* **2006**, *81*, 360–370.

78. Evans, A.E.V.; Hanjra, M.A.; Jiang, Y.; Qadir, M.; Drechsel, P. Water pollution in Asia: The urgent need for prevention and monitoring. *Water Qual.* **2012**, *9*, 1–4.

79. *State of the Environment in Asia and the Pacific*; United Nations Economic and Social Commission for Asia and the Pacific: Bangkok, Thailand, 2000.

80. Global Water Resources and Water for Agricultural Use in Japan. Available online: http://www.maff.go.jp/j/nousin/keityo/mizu_sigen/pdf/mizusigen_e.pdf (accessed on 20 April 2016).

12
CHAPTER

Issues and Challenges in Spatial and Temporal Water Allocation in the Nile Delta

Wouter Wolters [1], Robert Smit [2], Mohamed Nour El-Din [3], Eman Sayed Ahmed [4], Jochen Froebrich [1] and Henk Ritzema [5]

[1]*Alterra, Wageningen University and Research, P.O. Box 47, 6700 AA Wageningen, The Netherlands*

[2]*Central Directorate for Water Resources, National Water Resources Unit, c/o Ministry of Water resources and Irrigation, Corniche El Nil, Imbaba, 12666 Giza, Egypt*

[3]*Irrigation & Hydraulics Department of the Faculty of Engineering, Ain Shams University, 1 El Sarayat Street, Abbasseya, 11566 Cairo, Egypt*

[4]*Planning Sector, Ministry of Water Resources and Irrigation, Imbaba, 12666 Giza, Egypt*

[5]*Water Resources Management Group, Wageningen University, P.O. Box 47, 6700 AA Wageningen, The Netherlands*

ABSTRACT

"Egypt is a gift of the Nile," wrote Herodotus, and indeed, without the Nile there would be no Egypt as the world knows it. Egypt is mainly dependent on the flow in the Nile River (with an agreed share of 55.5 BCM) and it receives about 1.3 BCM rainfall annually (mainly along the north coast). The overall water use efficiency is already high, due to e.g., water scarcity and reuse of drainage water. Egypt's water resources are managed by the Ministry of Water Resources and Irrigation (MWRI), with agriculture as by far the largest user of irrigation water, with a share of about 85%. The purpose of the paper is to discuss the major issues and challenges in the spatial and temporal allocation of water, in relation to a free cropping pattern and the characteristics of the irrigation system. We conclude that the current world-wide call for "crop-demand-based precision irrigation supply" will not be easily attainable in Egypt. Instead, "water security" in the form of "guaranteed or agreed" water supply may be a preferred water allocation aim for various reasons, including lack of large storage possibilities, impossibility of fine-tuning supplies in the system, and the needed capacity to deal with (future) droughts. Although the paper concentrates on technical issues, it is increasingly realized that the challenges are not only of a technical nature and that there is a need for integration of policies as well as a need to establish effective science–business–policy interfaces at the national level.

Keywords: irrigation system; spatial and temporal water allocation; crop-demand-based water supply; drought; agriculture; integration of policies; water scarcity

1. INTRODUCTION

1.1. Egypt Is a Gift of the Nile

The Nile River Basin is, like the Indus basin, one of the oldest agricultural areas in the world [1] Egypt is an arid country with abundant sunshine and fertile soils, but rainfall is very scarce, thus agriculture has always depended upon irrigation. Average annual rainfall ranges from 1.5 mm in the south (near Aswan, about 900 km south of Cairo) to 150 mm in the north (in the coastal regions bordering the Mediterranean Sea, about 150 km north of Cairo. However, Egypt is blessed with the River Nile that provides 94% of the total water resources and 97% of the renewable water resources. There are limited quantities of water from rainfall and deep groundwater. These deep groundwater resources are mostly non-renewable and must therefore be used prudently with a long-term planning horizon.

The water provided by the Nile, combined with fertile soil and favorable climate, enables farmers to achieve relatively high crop yields. The agricultural sector provides employment to 32% of Egypt's work force and contributes about 14% of GDP to Egypt's economy [2].

Egypt's dependence on the river Nile is also illustrated by the fact that about 90% of the population lives on 5% of the land area around the Nile River and Delta. The remaining 95% of the country is desert area. The natural flow of the Nile has its annual peak in the summer and is subject to considerable variations over the years, depending mainly on the rainfall at the Ethiopian plateau. These variations tend to come in successive years of high or low floods. These multi-year cycles of high and low river flows were reason for Egypt to build the Aswan reservoir with sufficient storage to hold the excess water of high-yield years and use it in low-yield years.

1.2. Egypt's Water Resources Management/Development and Planning

From the days of the Pharaohs until the 19th century, basin irrigation was practiced. For this ancient method, based on the natural regime of the Nile, the natural drainage capacity of the land was sufficient to protect the area against the twin problem of waterlogging and salinity. In the 19th century, new crops,*i.e.*, cotton and sugarcane, were introduced that required water when the Nile's natural water levels were low. This resulted in the construction of barrages in the River Nile and a network of irrigation canals and open drains. The design was based on the British design school, based on the principle that the operation of such a large system had to be simple, run by as few people as possible and at the lowest cost [3]). With the construction of barrages in the River Nile and a network of irrigation canals, a continuous (day and night) fixed flow to a known area without the need for regular adjustments was achieved by gravity. Farmers were expected to handle their ration efficiently whenever it came, even at night. Adapting water gifts to changing demands in order to increase harvest was not done. With the completion of the High Aswan Dam in 1970, the Nile flow is now completely controlled. The High Aswan Dam (HAD) also enabled Egyptian farmers to move from seasonal to year-round irrigated cropping. Nowadays, the main crops are cotton, sugarcane, and paddy in the summer and wheat and berseem (Egyptian clover) in the winter. Land holdings are small, with about 78% less than 2.0 ha [4]. The year-round irrigation boosted agricultural production, but also made it necessary to install subsurface drainage systems to prevent salinization of the soil. Since the 1970s, the Egyptian Public Authority for Drainage Projects (EPADP, part of the Ministry of Water Resources and Irrigation) has installed almost 3.4 Mha with subsurface drainage systems [5,6]. The first dam near Aswan was built in 1903. The more recent planning of Egypt's water resources is based on the 1959 Nile Water Agreement, which allocates 55.5 billion cubic meters per year to Egypt. However, this international water agreement is no longer recognized by all riparian countries, some of whom demand a larger share of the Nile water to develop their economies.

The water balance of the country shows that agriculture is by far the largest consumer of water, with about 84% of the available water resources. The drinking water sector consumes 12% and industries consume 4%. To make up for water shortages, drainage water is recycled either directly by pumping it back into the Nile, into irrigation canals, or by pumping shallow groundwater in the Nile valley and delta. Reuse is practiced at the farm level, project level, and regional level, often as a combination of official reuse by the government and unofficial reuse by individual farmers. All these reuse options have increased Egypt's water availability by 20% [7]. At present only 22% of the fresh Nile water released from HAD reaches the Mediterranean Sea as drainage water. Together with water losses through evaporation from Lake Nasser, which are estimated to range from 10 to 15 BM^3 per year [8,9], the total water losses are about 26%. The overall water use efficiency is very high (about 80%), especially as we consider that drainage outflow to the sea is close to the minimum amount necessary to prevent saline water intrusion from the Mediterranean Sea and to sustain soil fertility of the Northern lands of the Nile Delta [10]. There is a general belief that through irrigation modernization large amounts of water can be saved through a switch to more efficient irrigation systems and infrastructure [11].

The purpose of this paper is to highlight the current major issues and challenges in the spatial and temporal water allocation in the Nile Delta. These issues and challenges are especially important in light of the free cropping pattern in Egypt and the characteristics of the irrigation system. The cropping pattern used to be dictated by the government but now farmers are free to grow the crops they like, which obviously gives rise to allocation issues. In addition, the characteristics of the system as well as the very limited possibilities to carefully adjust water supplies do not allow "crop-demand-based precision irrigation supply." We will also briefly compare with other "typical" irrigation systems in the world.

It is increasingly realized that the challenges are not only of a technical nature. The issue of water scarcity has consequences and implications that can no longer be adequately addressed by any one of the Ministries alone. There is a need for integration of policies and a need for trans-disciplinary approaches that should evolve in a "Water Scarcity Action Plan" as well as an effective science–business–policy interface at the national level. Food is grown by the private sector (including large farming organizations and small farmers) and this important stakeholder has a prime position in the Egyptian irrigation system (and its new extensions).

2. MAJOR OVERALL CHALLENGES FOR EGYPT

2.1. An Ever-increasing Pressure on Egypt's Water Resources

For decades Egypt's population grew by 2% annually; it has now reached approximately 90 million people. The fresh water availability per capita is closely related to the population growth and has decreased, as shown in the following Table 1:

Table 1. Trend of per capita fresh water availability.

Year	Population	Nile Water Availability
1960	25 million	2200 m^3/capita/year
2000	62 million	887 m^3/capita/year
2010	80 million	688 m^3/capita/year
2015	90 million	611 m^3/capita/year
2037	140 million	392 m^3/capita/year
2050	170 million	324 m^3/capita/year

Source: [12] (National Water Resources Plan 1997–2017).

Due to its population growth, Egypt is increasingly becoming a water-scarce country. Egypt's present population is approximately 90 million people, each of whom has a water share of 616 cubic meters per year, which is below the UN water-poverty criterion of 1000 cubic meters per capita per year. In order to feed its growing population, Egypt needs to boost its agricultural production [13], which can be done by e.g., developing non-conventional water resources (including reuse of drainage water, use of brackish water, desalination of sea water, and water harvesting in wadis) and increasing water productivity, by increasing yields per unit of water used (more crop per drop) and increasing the value of the water used (more cash per splash). Importing food is another way of solving the water shortage (virtually increasing the water supply) [14].

2.2. Water Pollution

Pollution of the surface water by agricultural drainage water, industrial wastewater and domestic solid waste and wastewater is a serious threat to the quality of irrigation and potable water supply [15]. While 99% of households are connected to a potable water supply system and 80% of the urban population is connected to public sewerage systems, in rural areas sewerage and solid waste services are largely absent. Only 26% of rural households are connected to public sewerage systems. The poor surface water and shallow groundwater quality in the Nile Delta forms a serious health risk through the reuse of drainage water and pumping of contaminated groundwater. It also frustrates government plans for downstream reuse of drainage water, which is a key element in its strategy to cope with water scarcity.

The National Water Resources Plan [12] formulates the key questions on water quality, health, and environmental aspects as follows: (i) what is the best mix of prevention, treatment and protection measures that results in a water quality that complies with reasonable standards?; and (ii) what is the level of investment needed to provide all people with safe drinking water and adequate sanitation facilities?

2.3. Effect of Climate Change

Studies show that the Middle East and North Africa (MENA) region will be one of the areas most affected by climate change. For Egypt this means that: (i) the low-lying Nile Delta will be threatened by seawater rise [16]; (ii) more extreme fluctuations in the Nile flows are anticipated [17]; and (iii) there will be an increased water requirement due to higher temperatures [18]. To date, there is no certainty about whether the Nile flow to Egypt may decrease or increase. The Egyptian share of the Nile flow of 55.5 BCM is about 5% of the 1660 BCM total rainfall in the entire Nile Basin. With 85% of the Nile flow originating in the Ethiopian highlands, it is clear that the future Nile flows depend for a large part on the impact of climate change on the rainfall in Ethiopia.

The overall sensitivity of the Nile River flow in response to changes in rainfall is considerable: if throughout the Basin rainfall were to decrease by 10%, inflow into Lake Nasser would decrease by more than 30%. Similarly, a small increase in rainfall will result in a significant increase in inflow [18]. The sensitivity of the Nile Basin to changes in rainfall is also observed in the historical data. For example, during the early 1980s the flow into Lake Nasser was 40% lower than during the late 1990s, while the difference in rainfall between the two periods was only 10%.

2.4. Effect of Urbanization and Infrastructure in the Delta

With some 50% of the Egyptian population living in the Nile Delta, the delta comprises many cities and towns. Furthermore, industry is a growing sector in the national economy. Further industrial development is expected to play a major role in the socioeconomic development of the country, providing employment for a large part of the growing population. Although the industrial policy is to create new cities and industrial zones outside of the Nile Valley and Delta, there is still a lot of industrial development in the delta as well. Moreover, through a rapidly spreading informal urbanization [19], there is increasing land fragmentation [20], which has an indirect effect on the water system. The extensive urban network of cities, villages, and connecting roads influences the irrigation networks due to excessive fragmentation. In addition, the increasing number of people living in urban environments in the delta are less aware of their huge and continuing dependence on water. It is also true that urban environments require less water than agricultural land. In summary, there is enormous pressure on both the quality and quantity of water in the delta, resulting from intensifying agriculture, rapidly growing industry, and a rapidly growing population. It is sometimes argued that if the current trends continue, the Delta will be a predominantly urban environment in 50 years from now.

2.5. Current Water Policy to Face the Challenges

The main challenge will be deciding how Egypt can safeguard its water resources in the future under the conditions of a growing population and more or less fixed water availability. Assuming that all available additional resources will be developed, the main questions with respect to water quantity that have to be answered are [12]:

- How can the efficiency of the various uses be increased?
- How can the agricultural expansion policies of the government be supported and what are the priorities and limitations in this expansion, given the existing water resources, optimum efficiency, and the priority for drinking and industrial water use?
- How should Egypt manage its water resources system under variable supply conditions?

Egypt's National Water Resources Plan [12] is based on a strategy that has been called "Facing the Challenge." This includes measures to

1) develop additional resources,
2) make better use of existing resources, and
3) measure water quality and environmental protection in the field.

There are various large programs ongoing in the Nile Delta to "face the challenge," including the Integrated Irrigation Management Improvement and Management Project (IIIMP); Enhanced Water Resources Management Project (EWRMP); the Second National Drainage Project (NDP2); the Integrated Sanitation and Sewerage Infrastructure Project (ISSIP); Improved Water Supply and Sanitation Services (IWSP); and the On-Farm Irrigation Development Project in the Old Lands (OFIDO).

2.6. Need for Integration of Policies

Future water allocation will have to be done keeping all water-using sectors in mind. The overall priority sequence is: first drinking water, then industry, and whatever is remaining will be available for agriculture and nature. The main option available in the short term to reduce water scarcity in the priority sectors of the economy is to allocate less to the agriculture sector [21]. However, the issue of water scarcity has consequences and implications that can no longer be adequately addressed by any one of the Ministries alone. Many other government departments and agencies must be involved and decisions will have to be made at the highest political level [22].

All policies in Egypt must be conscious of the severe limitations in water availability, and water policies need to address technological developments as well as the full range of other issues, including: macro-economic factors, economic issues that influence farm-level decisions, development of human capital, governance, and financial risk management. The government will be well equipped to deal with future uncertainties if it prepares a "scarcity outlook" that addresses the situation in 2050, and implements a "Water Scarcity Action Plan" [22] that will incorporate the knowledge that reform decisions are inherently political, that non-water policies are central to water management, and that improved accountability of government agencies and water service providers to the public is essential.

3. ISSUES AND CHALLENGES IN SPATIAL AND TEMPORAL WATER ALLOCATION

3.1. Aggravation of the "Seven Fat Years and Seven Lean Years" Phenomenon

The "seven fat years and seven lean years" phenomenon refers to the cyclic rhythm of the Nile flows in existence for thousands of years. The construction of the High Aswan Dam dampened this effect for Egypt, but the inflow of Lake Nasser is still subject to this variation and it needs to be mentioned that the construction of dams upstream in the Nile basin is expected to have a big impact on the flow to Egypt [23] and consequently also to the Nile Delta. The last major drought (period of lean years) was in the mid-1980s, but in the future droughts are predicted to become more severe, not only because of climate change but also because of the development of water resources upstream in the Nile Basin.

3.2. Description of the Irrigation System in the Nile Delta

As part of the overall system fed from the High Aswan Dam, the main canals operate essentially under a system of upstream control, with each irrigation directorate receiving a seasonally varied supply determined centrally by the Irrigation Sector in accordance with the expected cropping pattern. Releases from High Aswan Dam are made according to the predicted water requirements and taking into account the considerable travel time (measured in weeks) from Aswan to the Delta.

Control of discharges in the main canals takes place at the head regulators and at cross regulators located at the boundaries between irrigation directorates. The regulators are mostly equipped with lifting gates. Discharge control depends on rating curves either for the gates or for the downstream channel; these ratings are confirmed by periodic calibration through current metering. The method of operation emphasizes regulation to achieve specified water levels downstream of the regulators (as a proxy for discharge), rather than to maintain particular upstream water levels [1]. Upstream water levels at cross regulators may vary depending both on the season and on which branch canals are being fed from the upstream reach at any particular time under the rotation schedule: operation of the traditional irrigation system is based on rotational water deliveries to individual branch canals. The main feeder canals flow continuously and the off-taking branch canal head regulators are opened according to a rotation schedule, consisting of, e.g., a three-turn rotation or five days on/10 days off in winter and a two-turn rotation or five days on/five days off in summer. Because farmers lack confidence in the upstream control-rotational irrigation system they tend to over-irrigate, particularly head-end farmers, as an insurance against the uncertainty of the irrigation water supply [24]. Thus, they reduce water availability to tail-end farmers who, in turn, adopt a coping strategy that involves the reuse of drainage water even if it is saline (so-called "unofficial reuse").

Branch canal head regulators are generally also equipped with lifting gates, and again regulation of water deliveries is achieved by adjusting gates to maintain target downstream water levels. These target levels vary seasonally and serve as an indirect and rather approximate means of controlling discharges; in practice, they are determined to a large extent on the basis of accumulated experience.

3.3. Management of Water in the Nile Delta

In the absence of explicit measurement of discharge, the District Engineer, with the gate operator (bahar) who does the day-to-day control of the regulator gate, can exercise a certain amount of flexibility and discretion in allocating water supplies [1]. Sometimes, gate opening and rotation schedules may be adjusted in response to representations and complaints from farmers, or to observed conditions, e.g., the occurrence of escape flows at canal tails. In effect, the staff who are responsible for arranging water deliveries to individual branch canals try to distribute the water allocated to them in order to achieve the greatest possible degree of satisfaction of users. It is important to note that the irrigation system is supply-oriented at the upstream side, and demand-oriented at the downstream side. Obviously this is creating management issues.

It is also important to note that in managing water deliveries under the rotation system, operating staff are able to adjust not only the flow rate, but also the flow duration, by adding or subtracting days from the "on-period." This ability to use time as an operating variable allows operating staff to compensate for any surpluses or deficits arising from the imprecision of the discharge regulation, and in particular to deal with "emergency" shortages on individual canals. One adaptive practice that has been applied in recent years, especially during the peak period in rice areas, is the so-called "focused flow" [1], which involves concentrating the share of water from two or three canals into one canal at a time to ensure that water reaches the tail. Nevertheless, this cannot be done accurately as there is no volumetric control.

Downstream of branch canal offtakes, there is usually only limited hydraulic regulation. Except where they are needed for internal rotation or to control steps in water level, gates on cross regulators and sub-branch canal offtakes are often not functional or have been removed. Water levels depend primarily on abstraction of water by farmers.

3.4. Advantages of Improvements in Flow Regulation Being Tested

Due to certain negative effects of the rotational system, improvements are being tested [1]. These negative effects include: (i) non-flexibility in matching irrigation intervals and crop or soil condition; (ii) small effective channel storage capacity; (iii) high side slope deterioration during off periods and low bank stability; and (iv) the high cost of big hydraulic sizes of canal cross sections and structures.

The advantages of continuous flow compared to the rotation system are numerous in the opinion of both the users and the operators. The various advantages and their impacts found [1] in the tests of summer 2015 include, but are not limited to, (i) water availability at the tail of the canal is improved, and as an indicator of this both farmers and Irrigation Sector staff state that the number of complaints from tail users is drastically reduced; (ii) all parties emphasize that the irrigation practices can be adjusted to the water requirements of the crops rather than to the availability of water in the canal; (iii) another important advantage is that farmers in the tail reach can now develop irrigation strategies to reduce the salinity of the soil by reducing the reuse of drainage water and practicing leaching of saline soils through rice cultivation or in the period between two crops; and (iv) an advantage for water management is that farmers can use irrigation water efficiently. Under the rotation system farmers' priority was getting sufficient water, or more, because too much water is still better than too little water.

There are some cautions with these conclusions, because although the findings are encouraging (favoring continuous flow), these results may be misleading as (1) it was impossible to know the exact water volumes delivered to the canal during the whole season; and (2) the irrigation District did not receive complaints regarding water shortage during the 2015 summer season. This might indicate that supplied volumes of water to the canal were (too) generous.

3.5. The Mostly Unknown Role of Groundwater

A recent phenomenon of groundwater development and abstraction has been unfolding over the last few years largely under the radar of the water management staff in the Delta. Data about current groundwater abstraction levels in the Nile Delta are lacking and there are contradictions regarding the total irrigated area with groundwater in published data.

A recent study [25] found that: (1) groundwater use in the Nile Delta is a poorly studied phenomenon; few if any studies were concerned with the socioeconomic impact of using groundwater in old lands of the Nile Delta; (2) most of the wells are shared between various farmers: this shows that individual needs are still limited and seldom justify individual investment in a well (although this latter case is growing); (3) pumping from a well requires more energy than pumping from waterways, and costs on average 2 to 3 times more than pumping from the canal, with irrigation durations around twice as long too; (4) drilling and constructing a well costs around EP 11,000; and (5) these wells are in general only used intermittently and it is not always easy to assess the frequency of use by farmers. Farmers mostly use these wells in the summer (June and July).

Some farmers mentioned that they rely mainly on wells by a ratio of 70%–100% for irrigation in the summer season. Although conjunctive use may be seen as the key to flexibility and increasing supply to farmers, it is likely to increase sharply if overall supply has to be reduced, for whatever reason. The possible impacts of such a sharp rise would include: (i) a drop of dynamic water table level under 10 m; (ii) increased salinity locally; and (iii) increased saline groundwater intrusion.

4. DISCUSSION

In this section the issues and challenges are briefly discussed. It is increasingly realized that the challenges are not only of a technical nature.

4.1. No Real Possibility of Regulating Water Volumes

Discharge to main canals depends on rating curves either for the gates or for the downstream channel; these ratings are highly inaccurate (only confirmed by periodic current metering) [1]. This method attempts to achieve specified water levels downstream of the regulators as a proxy for discharge. For branch canals regulation of water deliveries is again achieved by adjusting gates to maintain target downstream water levels. These target levels vary seasonally and serve as an indirect and rather approximate means of controlling discharges; in practice, they are determined to a large extent on the basis of accumulated experience and this does not give insight into the volumes supplied.

4.2. Matching Irrigation Supply with Crop Demand Currently Impossible

The main canals operate essentially under a system of upstream control, with each irrigation directorate receiving a seasonally varied supply determined centrally by the Irrigation Sector in accordance with the expected cropping pattern. However, the farmers are free to choose their crops. The most basic requirements for crop-demand-based operation of an irrigation system include (1) the availability of variable quantities of water during the cropping seasons and (2) the possibility of, somehow, regulating the flows/volumes.

Tests to reduce the non-flexibility in matching irrigation intervals and crop or soil condition are being carried out in Egypt, but are hampered by the fact that the cropping pattern is not known and the flows/volumes cannot be properly measured and regulated. In practice this means that hopes for a better match between irrigation supplies and demand remain futile until a more accurate regulation of the available volumes is possible, provided the water is available. Again, the irrigation system is supply-oriented at the upstream side, and demand-oriented at the downstream side. As crop-demand-based precision irrigation supply will not be easily attainable in Egypt, providing water security in the form of guaranteed or agreed water supply may be a preferred water allocation principle. Proportional division of flow may have the best chances as the basis for water allocation in periods of drought. Instead of dividing water over the area, the water shortage is then divided over the land. There is a lack of awareness about the need to develop the capacity to deal with droughts in the future.

This issue in itself is not new to Egypt; it is documented for the Delta and also for Fayoum. Although the overall irrigation efficiency in Fayoum is relatively high, the irrigation water is not uniformly divided over the area [26]. Measurements of discharges and the division of flow over Fayoum showed that the improvement of irrigation water management should begin upstream, at the system's main flow-division structures. Once adequate water supplies to the lateral command areas have been secured, improvements further downstream in the system can be of value.

4.3. The Future Is Uncertain

The last major drought (period of "lean" years) was in the mid-1980s but the value of the assets in the Nile Valley and Delta has increased considerably since the 1980s and so has the population of Egypt. A drought of similar proportions can be expected to have more serious consequences nowadays. Climate scientists insist that future Nile flows depend for a large part on the impact of climate change on the rainfall in Ethiopia [18]. It is also known that droughts are predicted to become more severe in the future, not only because of climate change, but also because of development of water resources upstream in the Nile Basin. "Lean" years will aggravate the effect of upstream developments. This clearly shows the importance of Egypt being able to deal with (future) droughts. Changes in the cropping pattern recommended to farmers will be one of the tools to adapt to drier years.

4.4. The International Water System Dimension

The scale of the Egyptian water system is large and there are not many systems that are comparable in terms of size, importance, and type of users, *i.e.*, predominantly serving small land holdings.

An example of a comparable system is the irrigation system of Pakistan, which was developed by the same school of engineers as in Egypt [3]. In Pakistan, water managers cannot achieve crop-demand-based supply of irrigation water for various reasons [27]. Both the capacity of the canal system, at low rates of typically between 1.8 and 2 mm/day (much lower than in Egypt) and the availability of water are not sufficient for crop-demand-based supply of irrigation water, with the existing capacity of the reservoirs fully utilized. Lake Nasser in Egypt provides inter-annual storage, whereas the major reservoirs in Pakistan (such as Mangla and Tarbela) only provide intra-annual storage. There is monsoon rainfall in Pakistan, but the travel times in the system and the spatial distribution of monsoon rains are such that these two cannot be matched: the water for a field in the south of the Punjab province is released about three weeks before it actually reaches the field. The conclusion for the system in Pakistan was that it would be better to improve the performance of the present water allocation, thereby providing an assured or guaranteed water supply to water users, than to respond to field-generated demand.

A well-known example of an accurately managed system where water can be supplied when and where it is required is the Société du Canal de Provence et d'Aménagement de la Région Provençale (SCP), in France. This SCP was established in 1957, on the initiative of the French Ministry of Agriculture, by three local authorities: the administrative districts of the Var and Bouches-du-Rhône, and the City of Marseille [28]. Its mission is to develop and manage regional water resources, in order to provide a safe and reliable water supply for agriculture, industry and domestic use. The SCP is totally different from the Egyptian irrigation system in terms of institutional setup, size, and users. It is a public company, with 85% of its equity capital held by local authorities in the Provence-Alpes-Côte d'Azur region and 15% held by private partners (Chambers of agriculture and banks). It has about 200 km of main infrastructure including nearly 70 km of open canals and nearly 5000 km of supply and distribution pipes, with water coming from 85 dams and associated reservoirs. The relative abundance of rainfall and water storage capacity in the area combined with the possibilities of regulating the water flow accurately (because of the available head in the system) is the reason why water can be supplied when and where it is required.

4.5. The Need for Transdisciplinary Approaches on the Way to Sustainability

There is increasing insight that sustainable development requires fundamentally different approaches that do not only take into account the interaction of social, technical, and environmental systems [29], but also complexity and the uncertainties in decision-making as well as in the boundary conditions [30]. Trans-disciplinary approaches have emerged as powerful practices to shift from understanding sustainable challenges to developing socially robust orientations [31].

Understanding that uncertainty and complexity cannot be tackled by traditional deterministic methods, a stronger shared acceptance on possible solution strategies can be seen as the major driver to involve different sources of knowledge and extended peer communities (Spangenberg 2012). Of similar importance is the permanent process evaluation and co-learning during such a transdisciplinary approach [31].

Because there are many stakeholders in consuming and using water, it is clear that transdisciplinary approaches go far beyond traditional participatory approaches such as conducting interviews or farmer field days.

The implementation of such a transdisciplinary approach to improve the water management within the Nile Delta would primarily require a readiness to realize such a multi-stakeholder, multi-sector initiative, and a willingness to link effectively across different disciplines and policies. Food is grown by the private sector (large and small) and that sector needs to be involved as well. Secondly, it is important to realize that successful transdisciplinary approaches are often based in multi-annual pre-existing experience and a strong local facilitation capacity.

Understanding that there are a plethora of stakes in consuming and using the water of the River Nile, it is clear that implementing a true transdisciplinary approach in Egypt is not a trivial task and that it requires a serious commitment of all actors to cooperate over a longer period. To reach that, the presence of an adequate policy framework, institutional development, and the proper enabling environment is key.

In Egypt, policies and implementing institutions remain rather disconnected. There is a need for instruments that support the integration of policies and the integration of implementing steps. Without that, it will be difficult to foster cooperation between policy-makers and other stakeholders, including the food-producing private sector. Therefore, there is a need for effective science–policy–business interfaces at the national level. Understanding that all the past sector-oriented efforts have not led to final solutions and that there is not a really effective science–policy interface in operation may raise the expectation that using a transdisciplinary approach is worth trying. However, realizing the difficulty of integrating policies in Egypt, the sensitive political situation in the entire MENA region now and the limited local capacity to facilitate such a transdisciplinary approach suggests that for Egypt there is a need to tailor approaches specifically and to commit enough time to allow small success stories to grow.

5. CONCLUSIONS AND RECOMMENDATIONS

- There is enormous pressure on both the quality and quantity of water in the Delta, resulting from intensifying agriculture, rapidly growing industry, and a rapidly growing population. The rapidly spreading informal urbanization, leading to an extensive urban network of cities, villages, and connecting roads, influences the irrigation networks due to excessive fragmentation.
- Egypt is an arid country with abundant sunshine and fertile soils, but virtually without rainfall. However, Egypt is blessed with the River Nile, which provides 94% of total water resources and 97% of renewable water resources. The water provided by the Nile, combined with fertile soils and favorable climate, enables farmers to achieve relatively high crop yields.

- The last major drought (period of "lean" years) was in the mid-1980s, but in future the droughts are predicted to become more severe, not only because of climate change but also because of development of water resources upstream in the Nile Basin. Not only has the value of the assets in the Nile Valley and Delta increased considerably since the 1980s, so has the population of Egypt. A drought of similar proportions can be expected to have more serious consequences nowadays. These issues make it crucial for Egypt to be able to deal with (future) droughts and other climate-related extreme events: there is a need to develop capacity for assessment and management of water and climate-related risks in the future.
- Matching irrigation supply with crop demand is currently impossible because there is no real ability to properly measure and regulate water volumes at the distributary canal level. Moreover, releases from the High Aswan Dam are made according to crop water irrigation water requirements from an estimated cropping pattern, while the actual cropping pattern is free.
- As crop-demand-based precision irrigation supply will not be easily attainable in Egypt, providing water security in the form of guaranteed or agreed water supply may be a preferred water allocation principle. Proportional division of flow may have the best chances as the basis for water allocation in periods of drought. Instead of dividing water over the area, the water shortage is then divided over the land. There is a lack of awareness about the need to develop the capacity to deal with droughts in the future.
- The issue of (future) water scarcity has consequences and implications that can no longer be adequately addressed by any one of the Ministries alone. Many other government departments and agencies must be involved and decisions will have to be made at the highest political level, requiring the integration of policies. There is a need for a "Water Scarcity Action Plan" as well as effective science–business–policy interfaces at the national level. It is important that all involved realize that food is grown by the private sector (including large farming organizations and small farmers) and this important stakeholder has a prime position in the Egyptian irrigation system (and its new extensions).

ACKNOWLEDGMENTS

The authors would like to acknowledge the cooperation of many Egyptian and international water experts in their involvement in water management in Egypt, working in various projects over decades. The most recent projects include the various phases of the National Water Resources Plan projects, the Enhanced Water Resources Management Project (EWRMP), and the Egyptian–Dutch Advisory Panel Project (APP). Notwithstanding this, all opinions are those of the authors and do not necessarily represent the official views of any of those projects.

AUTHOR CONTRIBUTIONS

Wouter Wolters and Henk Ritzema conceived the idea of sharing the issues and challenges in water allocation in the Nile delta in this paper. They combined their experiences in Egypt and other countries with Mohamed Nour El-Din, Eman Sayed and Robert Smit, who are now involved, all in their respective positions, in water resources planning in Egypt and who all have an intricate knowledge of the current conditions of water management in the Delta. Jochen Froebrich shared his experience mainly in the sections on the international dimension of the paper and more specifically his vast and up-to-date knowledge in the area of "green economic growth" and science-business-policy interfaces.

CONFLICTS OF INTEREST

The authors declare no conflict of interest.

References

1. Enhanced Water Resources Management Project (EWRMP). *Continuous Flow Implementation in Nekla Command Area, Summer, Final Report*; MWRI: Cairo, Egypt, 2015.
2. World Bank Development Indicators. Available online: http://data.worldbank.org/data-catalog/world-development-indicators (accessed on 23 December 2015).
3. Ertsen, M.W. The Development of Irrigation Design Schools or How History Structures Human Action. *Irrig. Drain.* **2007**, *56*, 1–19.
4. Abdel Dayem, S.; Abdel Gawad, S.; Fahmy, H.; Vlotman, W.F.; Madramootoo, Ch.A. Drainage in Egypt: A story of determination, continuity, and success. *Irrig. Drain.* **2007**, *56* (S1), S101–S111.

5. Amer, M.H.; Abu-Zeid, M. History of Land Drainage in Egypt. In *Land Drainage in Egypt*; Amer, M.H., De Ridder, N.A., Eds.; Drainage Research Institute: Cairo, Egypt, 1989; pp. 43–66.

6. Fathi, M.; Hamza, A.M. Development of Land Drainage in Egypt and the Role of the Egyptian Public Authority for Drainage Projects. *Drainage along the River Nile*; Nijland, H.J., Ed.; Ministry of Public Works and Water Resources, Egypt, Ministry of Transport, Public Works and Water Management, Directorate-General of Public Works and Water Management: Lelystad, The Netherlands, 2000; pp. 21–51.

7. Barnes, J. Mixing waters: The reuse of agricultural drainage water in Egypt. *Geoforum* **2014**, *57*, 181–191.

8. El-Magd, I.H.A.; Ali, E. Estimation of the evaporative losses from Lake Nasser, Egypt using optical satellite imagery. *Int. J. Digit. Earth* **2012**, *5*, 133–146.

9. Elsawwaf, M.; Willems, P.; Pagano, A.; Berlamont, J. Evaporation estimates from Nasser Lake, Egypt, based on three floating station data and Bowen ratio energy budget. *Theor. Appl. Climatol.* **2010**, *100*, 439–465.

10. Ritzema, H.P.; Schultz, B. Optimizing subsurface drainage practices in irrigated agriculture in the semi-arid and arid regions: Experiences from Egypt, India and Pakistan. *Irrig. Drain.* **2011**, *60*, 360–369.

11. Fader, M.; Shi, S.; von Bloh, W.; Bondeau, A.; Cramer, W. Mediterranean irrigation under climate change: More efficient irrigation needed to compensate increases in irrigation water requirements. *Hydrol. Earth Syst. Sci. Discuss.* **2015**, *12*, 8459–8504.

12. MWRI. Water for the Future, National Water Resources Plan 2017. Available online: https://www.google.com/url?sa=t&rct=j&q=&esrc=s&source=web&cd=1&ved=0ahUKEwjb4-qIq5DMAhUCDSwKHU29CGgQFggcMAA&url=https%3a%2f%2fjkbharris.files.wordpress.com%2f2010%2f09%2fegypt-national-water-resources-plan-2017.pdf&usg=AFQjCNFEeiZ-RTPF5uKkupnPWFmzShJXrw&sig2=jIOSxoI0qaOYuePSvWUaHA&bvm=bv.119408272,d.bGg&cad=rja (accessed on 23 December 2015).

13. El-Agha, D.E.; Molden, D.J.; Ghanem, A.M. Performance assessment of irrigation water management in old lands of the Nile delta of Egypt. *Irrig. Drain. Syst.* **2011**, *25*, 215–236.

14. El-Gafy, I.K. System Dynamic Model for Crop Production, Water Footprint, and Virtual Water Nexus. *Water Resour. Manag.* **2014**, *28*, 4467–4490.

15. Abdel-Dayem, S. Water Quality Management in Egypt. *Water Resour. Dev.* **2011**, *27*, 181–202.

16. IDSC-UNDP. Egypt's National Strategy for Adaptation to Climate Change and Disaster Risk Reduction. Report by the IDSC, Information and Decision Support Centre of the Egyptian Cabinet and UNDP. Available online: http://faolex.fao.org/docs/pdf/egy141200.pdf (accessed on 8 April 2016).

17. Sušnik, J.; Vamvakeridou-Lyroudia, L.S.; Baumert, N.; Kloos, J.; Renaud, F.G.; La Jeunesse, I.; Mabrouk, B.; Savić, D.A.; Kapelan, Z.; Ludwig, R.; *et al.* Interdisciplinary assessment of sea-level rise and climate change impacts on the lower Nile delta, Egypt. *Sci. Total Environ.* **2015**, *503–504*, 279–288.

18. Ludwig, F.; Vellinga, P. *Impacts of Climate Change on Water Resource Management in Egypt and The Netherlands. Contribution to the 42nd Meeting of the Egyptian-Dutch Advisory Panel on Water Management*; MWRI: Cairo, Egypt; Alterra: Wageningen, The Netherlands, 2008.

19. Redeker, C.; Kantoush, S.A. The Nile Delta: Urbanizing on Diminishing Resources. *Built Environ.* **2014**, *40*, 201–212.

20. Alfiky, A.; Kauleb, G.; Salheena, M. Agricultural fragmentation of the Nile Delta; a modeling approach to measuring agricultural land deterioration in Egyptian Nile Delta. *Procedia Environ. Sci.* **2012**, *14*, 79–97.

21. World Bank. *Making the Most of Scarcity: Accountability for Better Water Management Results in the Middle East and North Africa*; IBRD: Washington, DC, USA, 2007.

22. El Guindy, S.; El Deen, M.S.; Wolters, W. Integrated policies are a must to face future water scarcity in Egypt. Available online: https://www.wageningenur.nl/de/Publicatie-details.htm?publicationId=publication-way-343230303735 (accessed on 8 April 2016).

23. Hammond, M. The Grand Ethiopian Renaissance Dam and the Blue Nile: Implications for Transboundary Water Governance. Available online: http://www.globalwaterforum.org/2013/02/18/the-grand-ethiopian-renaissance-dam-and-the-blue-nile-implications-for-transboundary-water-governance/ (accessed on 8 April 2016).

24. FAO-IPTRID. *Final Report: Rapid Assessment Study Towards Integrated Planning of Irrigation and Drainage in Egypt*; FAO: Rome, Italy, 2005.

25. El-Agha, D.; Closas, A.; Molle, F. *Survey of Groundwater Use in the Central Part of the Nile Delta, Water and Salt Management in the Nile Delta Project Report No. 6*; WMRI: Cairo, Egypt, 2015.

26. Wolters, W.; Ghobrial, N.S.; Bos, M.G. Division of irrigation water in The Fayoum, Egypt. *Irrig. Drain. Syst.* **1987**, *1*, 159–172.

27. Wolters, W.; Habib, Z.; Bhutta, M.N. Forget about crop-demand based canal irrigation in the plains of Pakistan. In Proceedings of Seminar on On-Farm. Salinity, Drainage, and Reclamation; Chaudhry, M.R., Kielen, N.C., Sufi, A.B., Eds.; WAPDA: Lahore, Pakistan, 1997; pp. 67–80.

28. SCP. Responsible Water Management for the Provence Region and Beyond. Available online: http://www.canal-de-provence.com/ (accessed on 23 December 2015).

29. Spangenberg, J.H. Sustainability science: A review, an analysis and some empirical lessons.*Environ. Conserv.* **2011**, *38*, 275–287.

30. Swart, R.; Raskin, P.; Robinson, J. Critical challenges for sustainability science. *Science (New York, NY)* **2002**, *297*, 1994–1995.

31. Scholz, R.W.; Steiner, G. Transdisciplinarity at the crossroads. *Sustain. Sci.* **2015**, *10*, 521–526.

13

CHAPTER

Geospatial Techniques for Improved Water Management in Jordan

Jawad T. Al-Bakri [1,*], Sari Shawash [2,†], Ali Ghanim [2,†] and Rania Abdelkhaleq [2,†]

[1]Department of Land, Water and Environment, Faculty of Agriculture, The University of Jordan, Amman 11942, Jordan

[2]The Ministry of Water and Irrigation, Amman 11181, Jordan

ABSTRACT

This research shows a case from Jordan where geospatial techniques were utilized for irrigation water auditing. The work was based on assessing records of groundwater abstraction in relation to irrigated areas and estimated crop water consumption in three water basins: Yarmouk, Amman-Zarqa and Azraq. Mapping of irrigated areas and crop water requirements was carried out using remote sensing data of Landsat 8 and daily weather records. The methodology was based on visual interpretation and the unsupervised classification for remote sensing data, supported by ground surveys. Net (NCWR) and gross (GCWR) crop water requirements were calculated by merging crop evapotranspiration (ETc), calculated from daily weather records, with maps of irrigated crops. Gross water requirements were compared with groundwater abstractions recorded at a farm level to assess the levels of abstraction in relation to groundwater safe yield. Results showed that irrigated area and GCWR were higher than officially recorded cropped area and abstracted groundwater. The over abstraction of groundwater was estimated to range from 144% to 360% of the safe yield in the three basins. Overlaying the maps of irrigation and groundwater wells enabled the Ministry of Water and Irrigation (MWI) to detect and uncover violations and illegal practices of irrigation, in the form of unlicensed wells, incorrect metering of pumped water and water conveyance for long distances. Results from the work were utilized at s high level of decision-making and changes to the water law were made, with remote sensing data being accredited for monitoring water resources in Jordan.

Keywords: remote sensing; Landsat 8 OLI; Rapid Eye; drylands; Jordan; water auditing

1. INTRODUCTION

Drylands are limited by soil moisture due to low rainfall and high evaporation. They show a gradient of increasing primary productivity in the order of hyper-arid, arid, and semiarid to dry sub-humid areas [1]. Drylands represent fragile ecosystems that are highly susceptible to environmental changes. Nevertheless, they provide important ecosystem services in terms of land and water, particularly in developing countries [2]. The limited surface water resources of drylands makes water management a challenge for planners and decision-makers, as groundwater will be pumped for different uses to compensate for the limited surface water. The indiscriminate use of groundwater, particularly for irrigation, may result in aquifer depletion and salinization of soil and water. The problems resulting from groundwater depletion are usually aggravated in countries with high population growth and limited water resources [3].

Among the countries of the Middle East, Jordan represents a challenging case in terms of dryland water management, as the country's water resources are limited and its population is growing at high rates [3]. In terms of water availability, Jordan is among the four poorest countries in the world, with a per capita share of less than 146 m^3 per year, which is far below the international water poverty line of 500 m^3 per year [4]. Drylands dominate most of Jordan, where more than 90% of the country's land is arid and receives low rainfall amounts (Figure 1). Therefore, agricultural production in the country depends on irrigation using both developed surface and groundwater resources. At present, agriculture consumes more than 60% of the developed water resources. This share is expected to decrease in the near future as the population is increasing at high rates that reached an average of 2.9% during 2000–2012 [3,5]. According to Jordan's Department of Statistics [6], the population of Jordan increased from 5 million in 2002 to 6.7 million in 2014. The problem of population growth during 2002–2012 was aggravated by the influx of 0.45 and 1.5 million refugees from Iraq and Syria, respectively [7,8]. Such conditions of scarce resources and high population would put more stresses on water use and management. In addition, they might create environmental threats resulting from the use of marginal water resources [9].

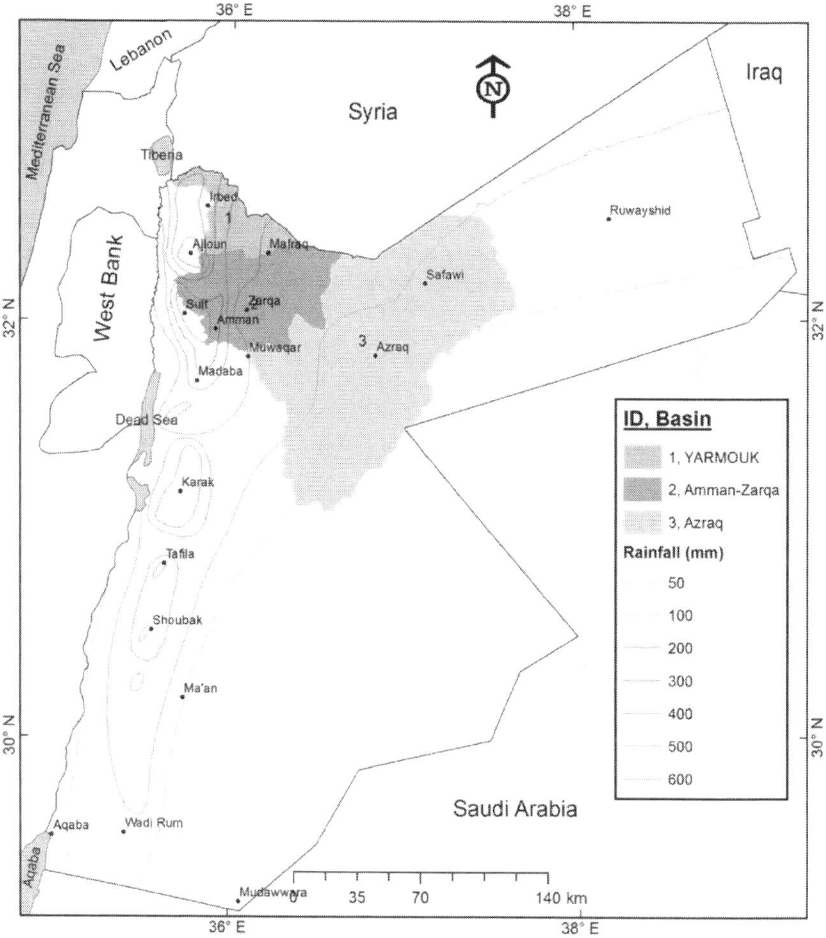

Figure 1. Location of the groundwater basins of Yarmouk, Amman-Zarqa and Azraq.

Managing water resources under the conditions of limited rainfall and increased competition among sectors requires updated information on water consumption by the different sectors. For domestic and industrial sectors, this can be achieved by the use of direct metering and information from pumping stations operated by the Government. For agricultural water use, the information on water consumption becomes more difficult when groundwater is used for irrigation. In Jordan, most of the figures on groundwater consumption for irrigation are inaccurate due to the lack of information on the actual irrigated area and crop water consumption. Information on the spatial distribution of irrigation with accurate areas for the different irrigated crops is still needed by the Ministry of Water and Irrigation (MWI) to assess and improve water management at the country level. The use of ground surveys to map irrigation is impractical, particularly when information is annually needed for water budget calculations and for revising plans of water management. Alternatively, remote sensing data and geospatial techniques can be used for this purpose. The improvements in spatial, spectral and temporal resolution of remote sensing data and the access to different sources of earth observation systems justify the adoption of geospatial techniques for managing water resources. The most important application in this regard is the use of medium and high resolution data to map crops and to estimate their evapotranspiration, the major component of crop water requirements or consumption use [10,11,12,13,14].

The use of remote sensing and geospatial techniques in water management took a new turn as these techniques were utilized in new approaches of assessment, generally called water accounting systems [15]. At the water basin level, the major components of these systems are the resource base, which represents water supply, crop evapotranspiration and water withdrawal, which represent water depletion, and the biomass production, which represents water productivity [16,17,18]. The most important inputs provided by remote sensing to the abovementioned components are the crop evapotranspiration and the land use maps [19]. The contribution of remote sensing and geospatial techniques will become more important in ungauged and poorly gauged basins as inputs will be provided with low cost and at reasonable accuracy [19]. Whether water accounting systems are implemented or not, remote sensing and geospatial techniques will continue to support water managers and decision makers with information needed for improved water management.

Within Jordan's efforts to manage its scarce water resources, MWI is in the stage of adopting geospatial techniques to improve water management at the country level. Recently, the MWI implemented a study in two pilot areas (Azraq and Mafraq) to map irrigated crops and to estimate agriculture water use using remote sensing data [20]. Initial results of the study showed that remote sensing maps were more accurate than estimates and records obtained by field crews of MWI. The difference between actual and estimated areas was important in terms of water consumption. Although the differences between both estimates could be attributed to the non-extensive ground survey that was limited by the available resources, they could, however, indicate possible violations of illegal drilling of wells and inaccurate metering of pumped water [21]. Therefore, the MWI adopted remote sensing for crop mapping and for auditing agricultural water consumption in the main irrigated areas in highlands. The areas included Jordan's parts of Yarmouk, Amman-Zarqa and Azraq basins (Figure 1). The overall aim of crop mapping using remote sensing data, presented in this work, is to provide MWI with information on the gap between official records and remote sensing estimates so that decisions can be taken to manage water resources in the three basins. Other objectives include the utilization of remote sensing data to map irrigated crops and the capacity building for MWI staff in using geospatial techniques for managing water resources. Despite their importance, spatial distribution of irrigated lands and their areas in Jordan were based on estimates, as the MWI lacked the capacity needed to fully utilize geospatial techniques. The inaccurate estimate of irrigated lands is also a worldwide problem and still remains uncertain [22]. Furthermore, the political nature of irrigation in countries that share water resources across borders often sets the stage for under-reporting of water use [23]. Therefore, this work provides proposed methods for mapping irrigated crops in Jordan's drylands and similar environments and will thus help in downscaling irrigation mapping to the basin level and provide important information that can be used for auditing records of water abstraction.

2. STUDY AREAS AND DATA COLLECTION

2.1. Study Areas

The first basin considered in this study was Yarmouk (Figure 1). About 30% (1393 km^2) of the basin's land is located in Jordan, while 70% is located in Syria. The basin has a semiarid Mediterranean climate in the west and an arid climate in the east. A rainfall gradient is obvious across the east-west direction, where the mean annual rainfall is 600 mm in the west and 150 mm in the east. The rainy season starts in November and ends by early May. The mean annual minimum and maximum temperatures in the west are 12.3 °C and 23.1 °C, respectively. The mean annual minimum and maximum temperatures in the east are 9.3 °C and 24.0 °C, respectively. The western parts of the basin are mainly rainfed areas cultivated with field crops, olives and vegetables, while the eastern parts extend to the low rainfall zone of the country where irrigation occurs using groundwater. Irrigation also occurs around Ramtha and in the northern parts of Jordan valley where citrus and vegetable crops are cultivated.

The second study area where irrigation was mapped is the Amman-Zarqa basin, also known as the Zarqa River basin. The basin has an area of 3600 km^2, where 95% of its area is inside Jordan and 5% is in the south of Syria. The basin has an arid climate in the east and the southeast, while the western parts have typical Mediterranean climates that are semiarid in Amman (Capital of Jordan) and dry sub-humid in Ajloun, where rainfall exceeds 560 mm. The western parts are mountainous and characterized by cool temperature in winter and mild temperature in summer. The annual rainfall ranges from more than 500 mm in the northwest to less than 100 mm in the east, with an average annual precipitation of 250 mm.

The third study area is Azraq basin (Figure 1), which has a total area 11,742 km^2. In this basin, water drains from all directions to the center (depression), which used to be a permanent oasis in Jordan's desert. The climate of Azraq Basin is arid with hot and dry conditions, leading to very high evapotranspiration rates. The mean annual temperature ranges between 7.4 and 24.5 °C in the west (Mafraq area) and between 2.6 to 36.6 °C in Azraq. The maximum air temperature in 2014 reached 42 °C. The mean annual rainfall is about 150 mm in the west and decreases to 70 mm in the east and south and to less than 50 mm in the center of the basin (Azraq depression). Over-pumping of groundwater for irrigation and drinking is noticed in this basin [24,25] and has resulted in the loss of wetlands surrounding the depression.

In terms of surface water resources, both the Yarmouk and Amman-Zarqa basins have perennial rivers that drain to Al-Wehdah and King Talal Dams (KTD), respectively. The collected water in both dams is used to irrigate different crops in Jordan Valley. Surface water resources in Azraq are mainly small desert dams that are used to irrigate small farms of olives and barley. Among the three basins, the Amman-Zarqa aquifer system is the most important groundwater supply in Jordan. This basin hosts more than half of Jordan's population [5] and more than 85 % of the industries in Jordan [26]. Groundwater resources in Yarmouk and Azraq basins, on the other hand, are also used for drinking and irrigation. Other sources of water include the treated wastewater from different plants in the three basins. Part of the effluent from treated wastewater plants (WWTP) is used for restricted agriculture, according to agreements between MWI and local farmers. The annual amount of treated wastewater effluent in Amman-Zarqa basin is 98 million cubic meters (MCM). About 15 MCM are used for irrigating forage crops in the basin, while the remaining is mixed with stored rainfall water in KTD and used to irrigate crops in Jordan Valley. The total amount of treated wastewater allocated for irrigation in Yarmouk is 3.8 MCM. Table 1 summarizes available water resources in the three basins and their use in 2014. All of the above figures were used in calculating gross groundwater amounts used for irrigation. In addition to over-pumping, other threats to water resources in the three basins are the trends of climate change, droughts and desertification [3,27,28,29].

Table 1. Summary of water resources in the study areas and their use in year 2014 (Source: MWI report of water budget [25]).

Basin	Groundwater Safe Yield (MCM)	Number of Agricultural Wells	Groundwater Use (MCM)		Surface Water Used in Irrigation (MCM)
			Agricultural	Non-Agricultural	
Yarmouk	40	129	36.4	9.4	3.8
Amman-Zarqa	88	590	63.9	92.4	15
Azraq	24	488	37.6	21.0	-

2.2. Remote Sensing Data

The main data used to map crop types were the images of the Operational Land Imager (OLI) of Landsat 8. The images have medium spatial resolution of 30 m. A higher resolution data of RapidEye, with 5-m pixel size, was used to delineate parcels of irrigated fields. The OLI data covered the period October 2013–September 2014, corresponding to a one-year period for calculating the water budget by the MWI. The data of RapidEye included two sets of cloud free images acquired in May and August 2014. Specifications of Landsat and RapidEye bands (B) used in the study are shown in Table 2.

Table 2. Specification of Landsat OLI and RapidEye images used in crop mapping for the three basins.

Satellite Data	Path/Row	Band (Wavelength in μm)	Use	Processing
Landsat 8 OLI	174/37 for Yarmouk, 174/37 and 174/38 for Amman-Zarqa, and 173/38 for Azraq.	B1 (0.43–0.45), B2 (0.45–0.51), B3 (0.53 –0.59), B4 (0.64–0.67), B5 (0.85–0.88), B6 (1.57–1.65), B7 (2.11–2.29).	Crop type identification	1. Atmospheric correction 2. Vegetation index
RapidEye	Mosaic datasets covering the three basins.	B1 (0.44–0.51), B2 (0.52–0.59), B3 (0.63 –0.68), B4 (0.69–0.73), B5 (0.76–0.85).	Delineation of fields	1. Geometric correction. 2. Visual interpretation

2.3. Ground Data

Ground surveys were conducted during May–October 2014 to collect data on the different irrigated crops in the three basins. During these surveys, information was collected on the types of crops, growth stage, planting and harvesting dates, in addition to irrigation systems and agricultural management followed for the different crops. This information was critical for verifying crop maps and for calculating water consumption for irrigated crops in the three basins. In total, field surveys covered 310 farms in the three basins.

Other ground data were collected for atmospheric correction of OLI images. Since remote sensing data of Landsat 8 were provided in digital numbers (DN) corresponding to radiance values at the top of atmosphere, it was important to convert the data into reflectance values at ground level. For this purpose, a handheld multispectral radiometer was used to collect ground measurements for target objects with standard or reference reflectance. The target objects were newly paved car parks (Dark objects) in front of Azraq Syrian Refugees Camp, dry desert mud flat, fields of parsley and tomatoes with full vegetation cover and bright surfaces (floors of abandoned quarries and desert surfaces). The selected locations were large enough so that they appeared in the form of several pixels on the images. Collection of ground data with the handheld radiometer was carried out within one hour of the satellite overpass on the 30 June 2014. The Landsat 8 image acquired for the day in which measurements were made was LC81730382014181LGN00.

2.4. MWI Data

Data of air temperature, relative humidity, wind speed and solar radiation were provided by the MWI. The data included daily records for the stations of Azraq, Safawi Um Ejjmal in Mafraq, Khirbet Samra, KTD, Amman, Ramtha and Irbed. The data were arranged in a spreadsheet and processed to interpolate missing records before being used for calculating crop water requirements. In addition, the MWI provided maps of groundwater wells, with attributes that included amounts of monthly abstraction, in addition to data on irrigated crops and crop management practices in Azraq and Amman-Zarqa basin. The data, originally collected from members of the highland water forum [30], were used for calculating crop water requirement and for verifying crop maps.

3. METHODOLOGY

Geospatial techniques and remote sensing data can be used for mapping irrigated lands at local, regional and global scales, according to objective, cost and time. The methods for crop mapping at local scales would include the visual interpretation or a digital classification method or a combination of both [10]. The improvement in remote sensing data encouraged researchers to implement and adopt digital classification methods, which are cost effective and consume less time when compared with the visual interpretation method [31]. The most commonly used approaches of digital classification include the parametric methods of supervised and unsupervised clustering [32]. These methods depend on the use of a single image acquired during the period of peak growth of crops. Due to multi-cropping and varying cropping calendars, a single image may not adequately characterize irrigated areas [33]. More advanced techniques may implement decision tree classifications, which are based on ground data, in addition to knowledge and experience of the classifier [10,34,35]. Further, the use of multi-temporal images of vegetation indices will increase the accuracy of mapping, as temporal resolution may compensate for the poor or moderate spatial resolution [10,36,37]. Among vegetation indices, the normalized difference vegetation index (NDVI) is considered a useful remote sensing product for mapping irrigation at local scales [36,37]. In this study, crop identification was based on the outputs from digital classification techniques of multi-temporal images of Landsat 8 (Figure 2), while boundaries of irrigated fields were digitized from very high resolution images of Google Earth (GE) and RapidEye (Figure 3), as described in the following subsections.

3.1. Processing of Remote Sensing Data

The set of downloaded images was processed to identify and exclude images with high cloud cover. A cloud mask was applied to each image using open source software [38]. Using the capabilities of a geographic information system (GIS), each image derived from the cloud mask was decoded using reclassification functions. The output images of cloud cover were displayed and processed in GIS to interpolate cloudy pixels in each image from previous and subsequent images. The total number of OLI images was 23 per year. However, the images with good quality and minimum cloud cover were only ten for Azraq and twelve for the two other basins. For the period of irrigation during April–September, most of images were free of clouds.

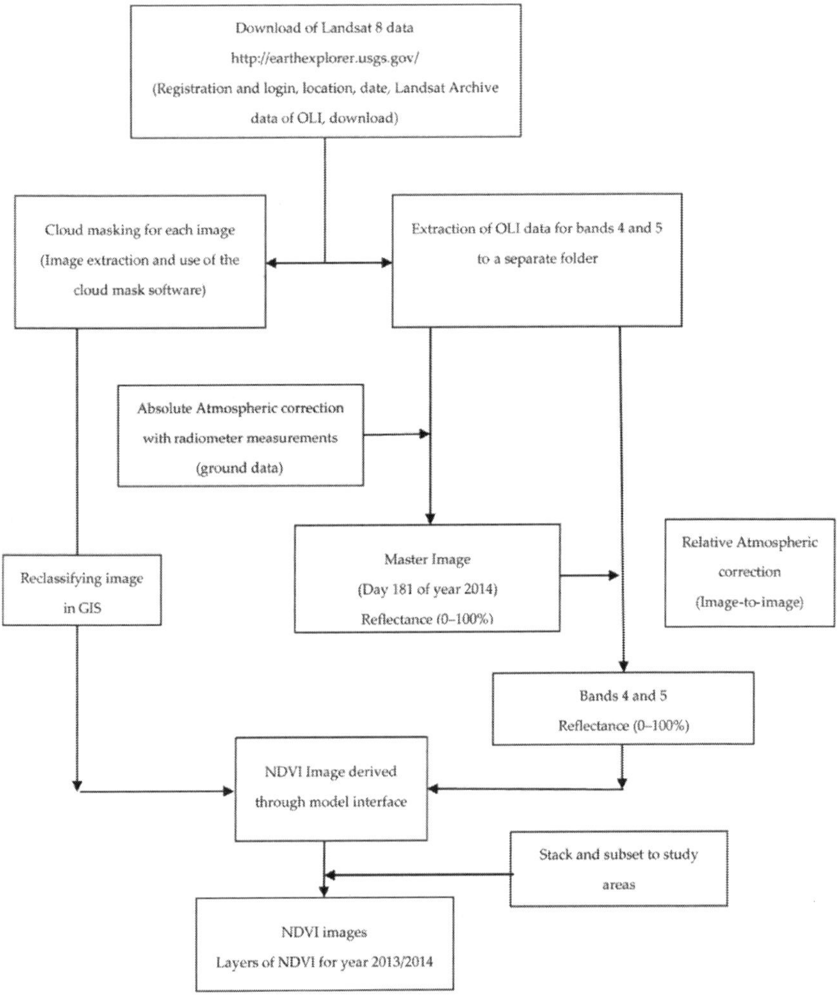

Figure 2. Flowchart of the image processing steps followed to derive NDVI images.

Figure 3. Use of high resolution images for digitizing parcels of irrigated lands and identifying ground cover for the different crops (Source: Google Earth).

Remote sensing data of OLI were corrected to obtain reflectance values at ground level. This was carried out using measurements from the handheld multispectral radiometer (with similar spectral regions to OLI). The data of the handheld radiometer were used for correcting one image (Day 181 of year 2014). This absolute atmospheric correction was based on empirical line calibration [39] derived from linear equations that correlated ground measurements of spectral reflectance with DN for locations of target objects that had unchangeable spectral reflectance among the set of OLI images. The multiple-date image normalization, using linear regression equations, was carried out to correct other images [40]. This relative atmospheric correction method was based on the use of invariant features that had unchangeable spectral reflectance among the set of OLI images. Both methods of absolute and relative atmospheric corrections were considered as standard methods that would not require information about atmospheric conditions at the time of image acquisition [41]. The correction was carried out for the red and near infrared bands of OLI. The corrected bands were then used to derive NDVI, based on the following equation:

$$NDVI = \frac{B5 - B4}{B5 + B4}, \quad (1)$$

where B4 and B5 are the red and near infrared bands, respectively. Both bands represent spectral reflectance at ground level. The steps of processing OLI images to derive NDVI are summarized inFigure 2.

3.2. Extraction and Analysis of NDVI Profiles

Observations collected during ground surveys were transferred into a GIS layer, representing points with different types of irrigated crops. The map was intersected with the layer of NDVI images to assign NDVI values for each crop type in the form of a profile that represented temporal changes in this index during the year. Examples of these profiles for Azraq are shown in Figure 4. The data of NDVI profiles were exported to spreadsheets and NDVI values were plotted for growing seasons. Statistical functions were applied in the spreadsheets to identify ranges of NDVI and to aid in building a decision tree for creating crop maps. Analysis of NDVI profiles showed that separation of the irrigated crops was possible with the use of multi-temporal images. NDVI profiles showed that alfalfa was characterized by the highest NDVI that reached its maximum theoretical value (1.0), with fluctuations resulting from crop cutting. Olives, on the other, had stable NDVI profiles (with a nearly constant value that was slightly above 0.50 for old plantations). Younger olives (3–4 years) had a straight line NDVI that was in the range of 0.20 to 0.30. The profiles for vegetables varied according to crop type and growing season, with peak NDVI values occurring during the midseason for each vegetable crop.

Following the stage of NDVI extraction and analysis, decisions were developed to identify the NDVI ranges (bounds) for each crop and for each time of the season. These decisions were mainly developed for Azraq and Amman-Azraq basins, while for Yarmouk, the unsupervised classification method was used to separate rainfed from irrigated areas. The NDVI range for each vegetable crop was based on the mean and standard deviation for NDVI image, as profiles of NDVI showed normal distribution patterns during the growing seasons. For alfalfa and olives, the range was modified according to the degree of agreement between classified pixels and ground observations. The NDVI profiles were also used to identify the length of the growing season, which was crucial for calculating crop evapotranspiration. Based on ground observations and analysis of NDVI, a classification scheme (Table 3) was adopted for mapping the different irrigated crops. The scheme included the different classes and sub-categories that included all irrigated crops and growing seasons in the three basins.

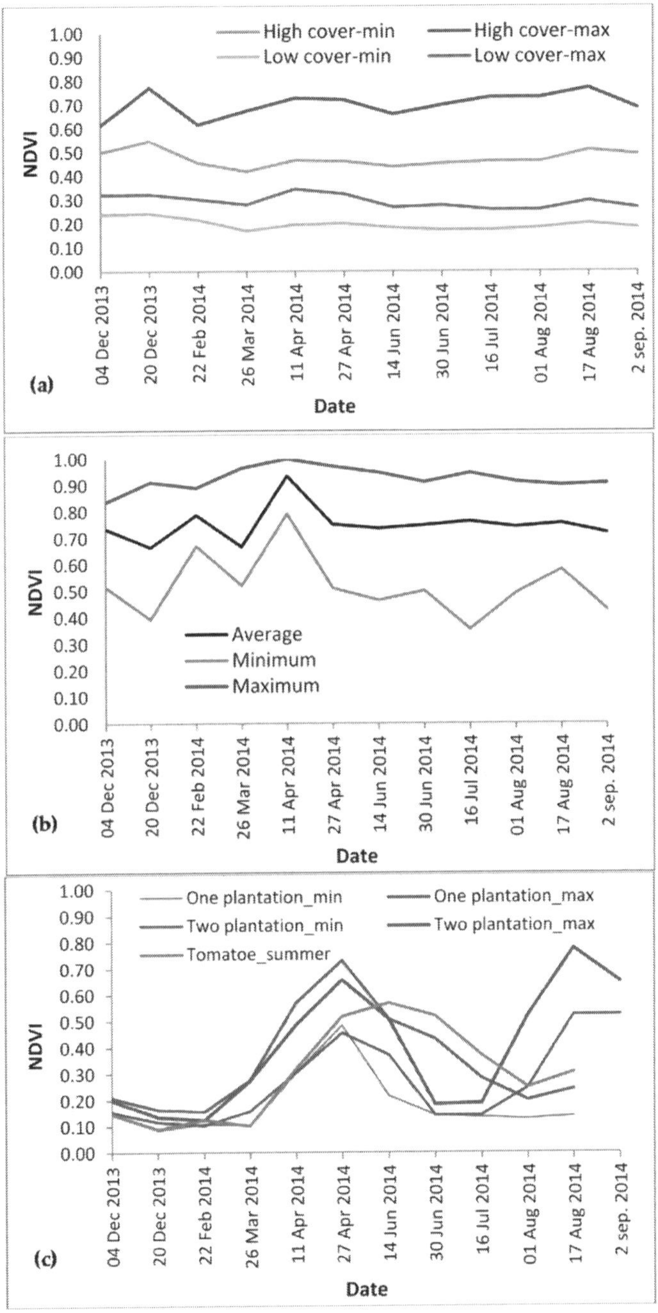

Figure 4. NDVI profiles of olives (**a**); Alfalfa (**b**) and vegetable crops (**c**) in Azraq area.

Table 3. Classification scheme of irrigated crops in Azraq and Mafraq.

Crop Category	Sub-Categories and Description
Olives	Farms of irrigated olive trees with different ground cover, depending on age and spacing. The class included three levels of cover that were grouped in the final map: low (<40%); medium (40%–60%); and high (>60%).
Fruit trees	Farms of deciduous orchards of peaches, apricots, stone fruits and table grapes grown on trellis. The sub-category of this class includes citrus in the northwest of Yarmouk.
Alfalfa and forage crops	Open spaces and farms cultivated with alfalfa under center pivot and solid sets of sprinkler irrigation systems in Azraq and fields of alfalfa and forage crops in the two other basins. The class includes barley fields cultivated under sprinkler irrigation.
Mixed cropping	Farms with two or more crop types, found mainly in Azraq. The fields include combinations of date palm, olives and alfalfa in the small landholdings near urban areas. The class is characterized by 100% ground cover for farms that included alfalfa in combination with olives, date palm and fruit trees.
Vegetables (Open fields)	Vegetables grown during February-June, March-July, April-August, June-September, July-October and September-December. The main irrigated crop is tomato. Other irrigated crops are melon, water melon, eggplant, zucchini, cauliflower, pepper and lettuce.
Vegetables and nursery plantations (Plastic houses)	Vegetables grown in plastic houses in the area of Mafraq and near Muwaqar in the middle west of the basin. The main crops are tomato and watermelon in Mafraq, and tomato and other vegetables in Muwaqar. The other sub-category of this class includes nurseries the lower part of Zarqa River between the WWTP and KTD and in the area of Baqa'a where some of the plastic houses are used as nurseries.

3.3. Mapping Irrigated Fields

Parcels of irrigated fields were mapped using an on-screen digitizing process with high resolution images of RapidEye and images of GE as backgrounds for delineation. The GE images were very useful as they had very high spatial resolution and horizontal accuracy that could reach 1 m [42]. RapidEye images, on the other hand, provided recent data that covered irrigated lands in year 2014, while GE images were mainly used in separating tree crops from vegetables. Identification of crop type, on the other hand, was based on outputs from the unsupervised classification of the multi-temporal images of Landsat 8 OLI. For Azraq and Amman-Zarqa basins, crop identification was carried out by the application of models that included decisions based on NDVI ranges derived from the multi-temporal images. The decisions were based on conditional statements that identified the range of NDVI for each crop and for each date in the set of the multi-temporal images of OLI. The models were applied in a sequence that prioritized crops according to their proportions, as observed during ground surveys. Subsequently, the sequence of models started with olives, followed by vegetables, fruit trees and other crops.

In Yarmouk basin, initial results showed some mixing between rainfed and irrigated areas. Therefore, principal component analysis (PCA) was used to transfer NDVI images of Yarmouk into three principal components. Applying the PCA to NDVI images resulted in transferring these images into three principal components that accounted for all seasonal changes in NDVI, as indicated by previous research [43]. Results from the unsupervised classification of PCA showed six distinguished spectral classes. The PCA images were then incorporated into unsupervised classifications with the ISODATA algorithm to derive spectral classes of irrigated and rainfed areas in this basin. Visual inspection of the output image showed that irrigated areas were distributed among three out of the six classes. In order to increase accuracy of mapping, models of knowledge based classifiers were used to refine the map of irrigated crops by incorporating layers of digital elevation model (DEM), rainfall and images of minimum NDVI values in these models. The DEM was used to exclude areas with high altitudes, where Jordan's forests were located. A rainfall map was used to exclude areas with high rainfall, as rainfed agriculture was dominant in these locations. The map of minimum NDVI separated irrigated areas from natural vegetation in the low rainfall zone. Accuracy assessment was made for the output maps using the confusion-matrix method [44], which compared remote sensing results with observations collected during ground surveys.

3.4. Assessment of Groundwater Abstraction Records

Assessment of groundwater abstraction for irrigation would require the use of ground measurements to compare amounts of abstraction with crop water consumption. Ground measurements of crop water requirements, however, are time consuming and become very expensive for large geographical areas. Therefore, crop water requirements can be calculated using crop maps and meteorological data. Remote sensing techniques can also provide spatial estimates of crop evapotranspiration (ETc), which is the main component of net crop water requirements (NCWR). This can be achieved by the surface energy balance models that utilize remote sensing data to derive the main factors controlling water evaporation from soil and plants and thus calculate ETc [45]. These models, however, require validation before being adopted for calculating ETc [46]. Therefore, crop maps were utilized in this study to calculate the crop ETc using the standard method of Food and Agricultural Organization [47]. The other component included in calculations of NCWR was the salt leaching requirement, which depended on water salinity and crop type. Rainfall in Yarmouk and Amman basin was also considered in these calculations. The FAO56, based on Penman-Monteith method, was used to calculate ETc as follows [47]:

$$\mathrm{ETc} = \mathrm{ETo} \times \mathrm{Kc}, \tag{2}$$

where ETc is the actual crop evapotranspiration (mm); Kc is the mean monthly crop coefficient; and ETo is the grass reference evapotranspiration. The NCWR, in million cubic meters (MCM), was calculated by multiplying the seasonal ET of the crop with its corresponding area, taking rainfall into consideration.

The FAO56 method calculates the theoretical crop water requirements and assumes standard conditions of crop spacing, ground cover and growing periods. These assumed conditions might not be valid in all fields in the three basins. Therefore, crop spacing and cover were considered in the calculations by adjusting Kc values based on ground cover fraction [48]. Ground cover was derived from the high resolution images of GE (Figure 3) and the data collected during ground surveys. The length of each growing stage was adjusted from NDVI profiles (Figure 4), which showed some differences between the values reported by the FAO56 method and the actual growing periods. Following this step, daily weather records of MWI were used to calculate ETo using an ETo calculator, developed by FAO [49]. The outputs of daily ETo were summed for each month to calculate monthly ETc. Values of Kc were adjusted using NDVI profiles and data collected during ground surveys. The adjustment was made according to ground cover and used a linear interpolation to derive monthly Kc. Irrigation efficiency was used to convert the NCWR to gross crop water requirements (GCWR). The average irrigation efficiency in the three basins was about 80% [50]. The calculated GCWR were compared with the records of groundwater abstraction for irrigation purposes to assess the compliance between the recorded abstraction in relation to crop water requirements and safe yields in the three basins. For vegetables cultivated in plastic houses and nurseries, the GCWR were given an average value of 800 m^3 per year, based on the average for tomato [51,52] for two cultivations per year.

4. RESULTS AND DISCUSSION

4.1. Ground Surveys

Observations from ground surveys showed that olives, fruit trees and vegetables were the main irrigated crops in the three basins. Irrigated vegetables included tomato as the main crop and other crops of melon, water melon, eggplant, zucchini, cauliflower and lettuce. Fruit trees included peaches, apricots, table grapes, pomegranate and date palm (in Azraq only). The size of some irrigated farms reached 400 ha (equivalent to 4000 dunums in the locally used unit), while the size of vegetable farms (open fields) was in the range of 20–30 ha and reached 70 ha in some areas, indicating considerable investment in irrigated agriculture. In the area of Mafraq in Amman-Zarqa basin, many of the drylands were turned into agricultural fields by clearing soil surface from basalt rocks.

This practice, shown inFigure 5, had been conducted since the 1980s, when irrigation started in the drylands of Mafraq [53]. Ground observations also showed different planting and harvesting dates for vegetables, which were confirmed by the NDVI profiles (Figure 4).

Figure 5. Examples on irrigated crops and irrigation practices in the three basins including fields of tomatoe under drip irrigation (**a**) and alfalfa under center pivot irrigation (**b**); watermelon at harvest stage (**c**); fields of cauliflower cultivated during March–July (**d**); an irrigated farm of table grapes (**e**) and a piece of land in Mafraq (inside Amman-Zarqa basin) with the soil surface partially cleared from basalt rocks to cultivate tomatoes (**f**).

4.2. Crop Maps and Irrigated Areas

4.2.1. Irrigation in Yarmouk

The crop map of Yarmouk showed that the total irrigated area was 5.9 thousand ha, distributed in different parts of this basin (Figure 6). The main irrigated crops were tree crops (olives and fruits) and vegetables. Analysis of the crop map (Table 4) showed that vegetables constituted about half of the irrigated area, while the other half included olives and fruit trees (48%), with a small proportion of forage crops and nurseries. Most irrigation was practiced in the drylands of Mafraq, followed by the north middle areas. In addition, irrigation was practiced in the western parts of Ramtha, where the total irrigated area in that part constituted 21% of irrigation taking place in the basin. Irrigation was limited in the southwestern parts, as the area was mountainous and lacked irrigation infrastructure (dams and groundwater wells). The absence of irrigation to the west and north of Irbid would be mainly attributed to the high rainfall that supported the rainfed agriculture, which included olives and field crops. Further, urbanization was encroaching into these areas and was competing with other land uses.

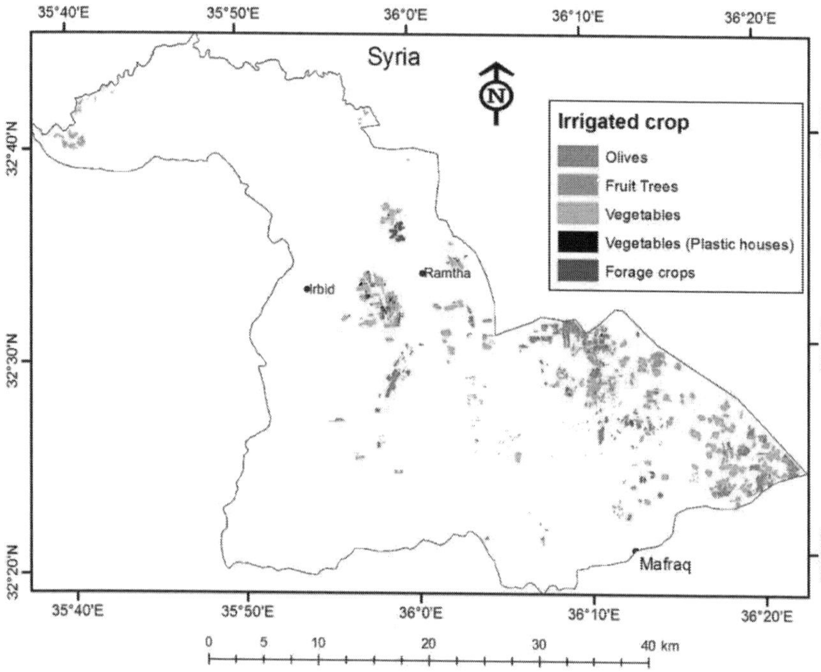

Figure 6. Distribution of irrigated crops in Yarmouk basin, as derived from remote sensing data.

Table 4. Analysis of irrigated areas in the three basins, as mapped from remote sensing data for the period October 2013–September 2014.

Class	Yarmouk		Amman-Zarqa		Azraq	
	Area (ha)	%	Area (ha)	%	Area (ha)	%
Olives	2018	34.1	4986	27.5	3050	39.0
Fruit trees	837	14.1	4118	22.7	1286	16.5
Alfalfa and forage crops	187	3.2	638	3.5	587	7.5
Mixed cropping	-	-	439	2.4	425	5.4
Vegetables (Open fields)	2763	46.7	7737	42.7	2465	31.6
Vegetables and nursery plantations (Plastic houses)	115	1.9	210	1.2	-	-
Total	**5920**		**18,128**		**7811**	

In terms of irrigated areas, this study provided detailed maps of irrigation in Yarmouk basin. Previous work [28,29] showed similar spatial distribution of irrigation in the middle and eastern parts of this basin without details on crop types. Therefore, this study contributed to the governmental efforts in managing water resources by providing more detailed maps for irrigation, particularly in the western parts of the basin. The best estimate of MWI for irrigated lands in the basin was 2.1 thousand ha, while initial results obtained from the digital classification of PCA images of NDVI showed that the irrigated area was more than 5.1 thousand ha [21,54]. The reason behind these counterintuitive findings could be the non-extensive ground survey carried out by MWI, which was limited by time and cost. Such results, therefore, encouraged MWI to take more actions toward the adoption of remote sensing technology for mapping irrigation in the basin and to audit data of groundwater abstraction.

4.2.2. Irrigation in Amman-Zarqa

The total irrigated area in Amman-Zarqa basin was 18.1 thousand ha. Vegetables constituted 43% of this area, while olives and fruit trees constituted more than half of the irrigated lands. Irrigation was practiced in two areas: Mafraq-Hallabat and on both sides of Zarqa River (Figure 7). This distribution could be attributed to the sources of irrigation water, which were groundwater in the Mafraq-Hallabat and the surface water of Zarqa River. The disposed treated wastewater was used for irrigating forage crops and tree plantations in nurseries, although some violations were observed on some farms on the sides of the river. Further analysis of the irrigation map showed that 77% of irrigation was taking place in Mafraq-Hallabat, indicating that groundwater was the main source for irrigating these drylands. The area to the north of KTD had limited irrigation, as the high rainfall encouraged rainfed agriculture in this mountainous area. In terms of irrigated areas in the basin, figures from this study were less than those reported by a previous study [26], which used unsupervised classification of a single image of Landsat ETM+ and reported a total irrigated area of 27 thousand ha. This could be attributed to the different crops cultivated during these periods and the time difference between this study and previous ones.

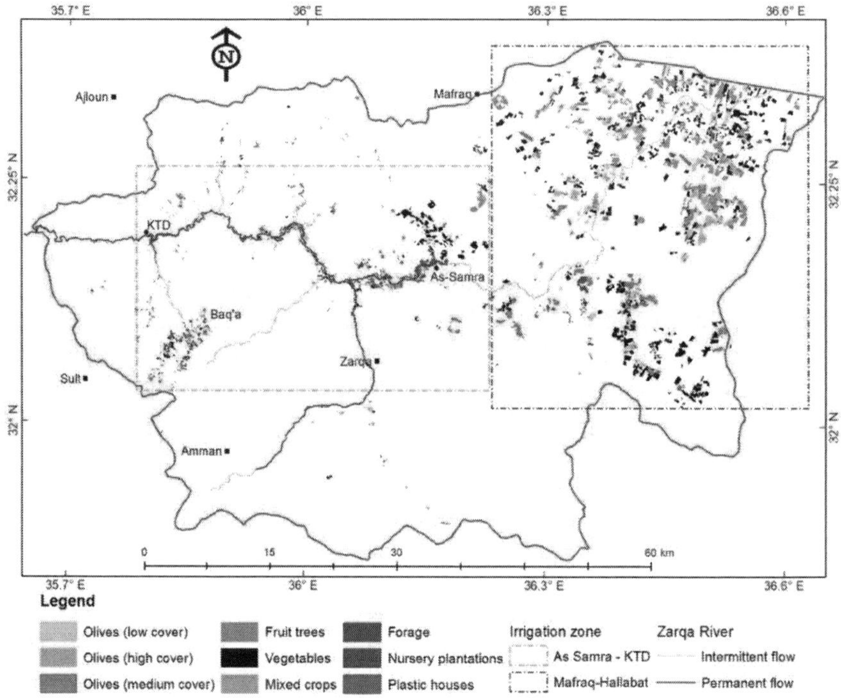

Figure 7. Map of irrigation in the Amman-Zarqa basin as derived from remote sensing data.

4.2.3. Irrigation in Azraq

Analysis of the irrigation map showed that the total irrigated area in Azraq basin was 7.8 thousand ha, mainly distributed in two strips in the basin, Azraq depression and northwest of Azraq (Figure 8). The reason behind this spatial distribution could be attributed to the characteristics of groundwater in these two parts, where the irrigated areas were located in the unconfined aquifer of quaternary basalt outcrop overlaying the aquifer A7/B2. The A7/B2 aquifer, which has two joined and fissured formations (Amman formation (B2) and the older Wadi Es-Sir formation (A7)), is characterized by high conductivity in the form of solution channels and karstic features and extends into the northern parts of Jordan [24,55]. The absence of irrigation in the southern parts of the basin would be attributed to the distribution of B4 (very low transmissivity aquifer) in these areas [24,55]. Analysis of the crop map showed that the main irrigated crops were olives (39%), followed by vegetables (32%) and fruit trees (17%), while other crops were alfalfa and mixed crops (Table 4). In terms of irrigated areas, figures from this study agreed with those obtained in previous work that used visual interpretation of satellite images [24]. On the other hand, an estimate that was based on administrative boundaries of the basin reported that the total irrigated area in Azraq would be in the range of 11.5 thousand ha [50]. These findings clearly indicated the improvement in mapping irrigated areas when multi-temporal remote sensing data were used.

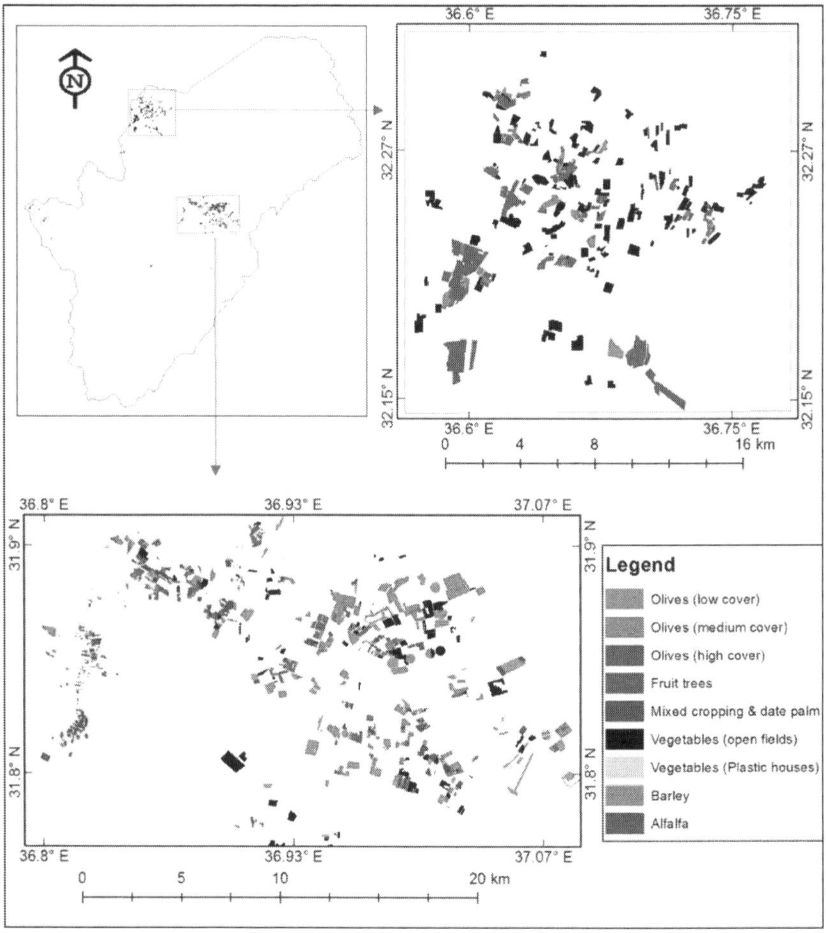

Figure 8. Map of irrigation in Azraq basin as derived from remote sensing data.

4.3. Mapping Accuracy

Although remote sensing data could provide important maps for irrigated lands, their accuracy should be assessed before being used. Comparing field observations (310 points) with maps of irrigated crops showed a good agreement between irrigation maps and ground data, with an overall accuracy of 87% (Table 5). The minimum accuracy was observed for fruit trees, which were mixed, in terms of classification, with olives. Considering the calculations of NCWR from ETc (Table 6), this would mean that NCWR (Table 7) for fruit trees was nearly 10% more than olives. For other irrigated crops, the level of mapping accuracy was relatively high and would be acceptable for the purpose of mapping and for estimating agricultural water consumption in the three basins.

Table 5. Analysis of irrigated areas in the three basins, as mapped from remote sensing data for the period October 2013–September 2014.

	Reference Data								
	Class	**Olives**	**Alfalfa & Forage**	**Fruit Trees**	**Vegetables**	**Barley**	**Mixed**	**Totals**	**Mapping Accuracy (%)**
Classified Image Data	Olives	89		6	2	-	-	97	92
	Alfalfa & Forage	2	32	-	-	-	-	34	94
	Fruit Trees	24	-	44	-	-	-	68	65
	Vegetables	-	1	-	75	-	3	79	95
	Barley	-		-	-	9	1	10	90
	Mixed	-	2	-	-		20	22	91
	Totals	**115**	**35**	**50**	**77**	**9**	**24**	**310**	**86.8**

Table 6. Monthly evapotranspiration for the main irrigated crops in Azraq area.

Month *	**ETo (mm)**	**Crop Evapotranspiration (ETc) in mm**								
		Olives	**Fruit Trees**	**Mixed**	**Alfalfa**	**Vegetables**				
October	104	67	67	80	97	-	-	-	-	77
November	62	39	25	45	55	-	-	-	-	-
December	36	10	–	15	41	-	-	-	-	-
January	45	15	–	10	43	-	-	-	-	-
February	65	22	–	28	75	26	-	-	-	-
March	102	62	49	72	88	72	41	-	-	-
April	158	102	110	122	147	158	110	63	-	-
May	176	102	116	129	157	101	176	123	-	-
June	203	119	183	158	190	41	117	203	81	-
July	242	149	218	188	226	–	48	139	162	97
August	237	130	190	184	221	–	–	47	213	158
September	156	96	109	121	145	–	–	–	130	140
Total	**1586**	**915**	**1067**	**1151**	**1484**	**397**	**493**	**576**	**586**	**472**

* Starting from November 2013 to September 2014, corresponding to one year of water budget at MWI.

Table 7. Summary of irrigated crops and their net annual water consumption.

Class	Yarmouk		Amman-Zarqa		Azraq	
	ETc * (mm)	NCWR (MCM)	ETc * (mm)	NCWR (MCM)	ETc * (mm)	NCWR (MCM)
Olives	700	14.1	726	36.2	887	27.0
Fruit trees	714	6.0	765	31.5	762	9.8
Alfalfa and forage crops	1213	2.3	1206	7.7	1320	7.8
Mixed cropping	-	-	888	3.9	1148	4.9
Vegetables (Open fields)	538	14.5	441	34.1	494	12.2
Vegetables and nursery plantations (Plastic houses)	800	0.9	800	1.7	-	–
Total		37.8		115.1		61.7

* Weighted average for all crop sub-categories.

4.4. Assessment of Groundwater Abstraction Records

One important tool for water management would be the use of outputs from crop mapping for water accounting and auditing. This was possible by deploying crop maps and weather records to calculate evapotranspiration, which is considered as the main component of net crop water requirements (NCWR). Results showed relatively high ETc values for forage, olives and tree crops in the three basins, particularly Azraq (Table 6). For vegetable crops, the NCWR varied according to cultivation season, which was accurately detected by the NDVI profiles. Although not investigated in this study, these profiles could have higher contribution to NCWR by deriving Kc with reasonable accuracy [56].

Variations in the ETc level among the three basins could be attributed to several factors including climatic conditions, the length of growing seasons and management practices that included different tree spacing in the three basins. Considering ETc and the area of each irrigated crop, the NCWR was calculated in the three basins (Table 7). Results showed that olives and fruit trees were the main consumers of water in the three basins. The NCWR for vegetables was variable according to planting and harvesting dates (Table 6). These results emphasized that cropping patterns in the three basins should be revised and changed to sustain groundwater resources. Proposed actions could include the prohibition of irrigating olives and alfalfa with fresh water resources of the three basins and the gradual replacement of both crops with other crops that consume less water. In addition, cultivation of vegetables during the summer season should be discouraged as it would increase water consumption by 30% (as implied from the ETc calculations). Another aspect of water management could be the use of flexible water allocation and pricing according to crop type and cropping season. A previous study in Jordan highlands showed that there was a potential to decrease water consumption and to reallocate it in an optimal way that considered cropping pattern and income from the irrigated area [57].

Outputs from crop mapping and NCWR calculations were used to assess groundwater abstraction records obtained by the MWI. The assessment was made by comparing MWI records with the NCWR, calculated by the FAO56 method after modifying Kc and the length of the growing season. This water auditing task was achieved by considering NCWR, irrigation efficiency and water resources used in irrigation to calculate GCWR. Results are summarized in Table 8. For the three basins, the GCWR were higher than the safe yield of groundwater. In Jordan, the term of groundwater safe yield is based on water balance models calibrated with data from groundwater monitoring wells [58].

Table 8. Groundwater abstraction records compared with estimates from remote sensing.

Basin	Safe Yield (MCM)	Groundwater Abstraction for Irrigation		Agricultural Abstraction/Safe Yield (%)	Abstraction/Safe Yield * (%)
		MWI Records	Remote Sensing		
Yarmouk	40	36.4	48	120	144
Amman-Zarqa	88	63.9	104	118	224
Azraq	24	37.6	67	279	367

* Including non-agricultural water uses (Table 1) and calculated as abstraction/safe yield.

In terms of agricultural water use, irrigation consumed considerable amounts of water that exceeded the safe yields in the three basins. Considering all uses of water, the over abstraction of groundwater was in the range of 144% in Yarmouk to 367% in Azraq. These results were in line with findings from the study on a groundwater aquifer in the north of Jordan [58], which indicated a decline of groundwater levels by 40–60 m during 1986–2014. This decline resulted in changing water flow directions from Zarqa towards the east of Mafraq and to the north in Yarmouk basin, *i.e.*, an opposite direction to its movement 40 years before [58]. Further, the results from Azraq indicated an over-abstraction of groundwater that could result in negative consequences resulting from groundwater depletion. The results for Azraq were also supported by findings from a previous study [59] that indicated a decline and lowering of the water table of the upper aquifer by 20 m during 1983–2003, the period during which irrigation had expanded in this basin.

Findings from crop mapping and GCWR calculations were alarming to MWI as they indicated that expansion in irrigation and adoption of inappropriate cropping patterns could result in future loss of groundwater resources. In addition, the shift between MWI records and remote sensing estimates indicated uncovered violations pertaining to groundwater use. Therefore, the work of crop mapping was extended to study the spatial distribution of irrigated areas in relation to water sources. The disagreement between water sources and irrigated areas was used to highlight areas with possible violations pertaining to water use or even possible water theft. The first map was prepared for the area around Ramtha City in Yarmouk basin (Red boxes in Figure 9).

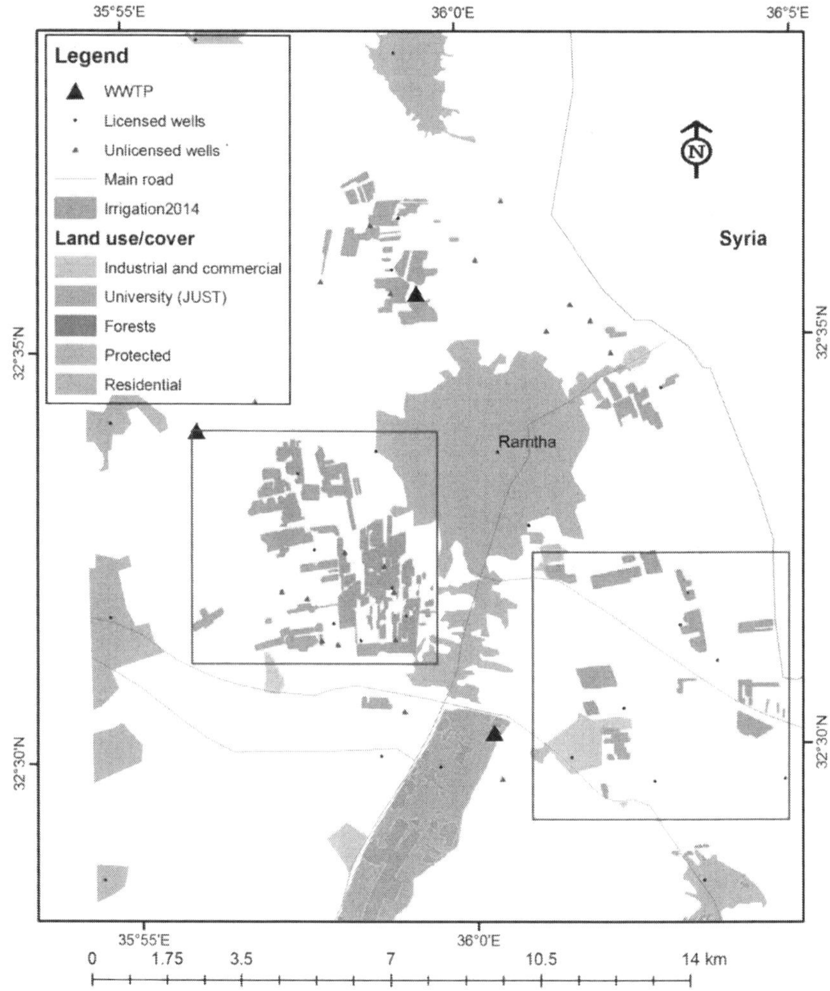

Figure 9. Map of irrigated areas and distribution water sources around Ramtha City.

Results from crop mapping showed that the irrigated area was nearly twice the area estimated by the field crew of MWI, while GCWR values were nearly three times higher than the recorded abstraction. Overlaying maps of irrigated crops and groundwater sources (Figure 9) showed that many irrigated fields were located in areas without groundwater wells or available surface water sources, indicating an illegal practice or violation to water use. In such cases, the concept of violation or illegal irrigation would be related to water pumping and conveyance over long distances rather than to cultivation of certain crops. Therefore, ground visits were carried out in fields that were located far away from water resources to uncover these violations, which were mainly in the form of water conveyance with pipelines for long distances or unlicensed groundwater wells [21]. In some areas, irrigation pipelines were illegally connected to the domestic water network and water was transferred to irrigate fields of vegetables. These findings urged MWI to carry out official campaigns to uncover these violations and to announce findings through mass media [54].

The outputs from this work were reflected in MWI decisions to improve water management and to activate water accounting and auditing. At present, amendments made to the "Water Law" state that satellite images and remote sensing techniques are officially adopted for auditing and estimating amounts of abstracted groundwater through crop mapping and estimation of water requirements for the different irrigated crops [60]. According to Article 4 of the new "Water Law", satellite images are among the accredited means that can be used by MWI to map irrigated areas and crop type and to estimate amounts of groundwater abstraction for irrigated farms. Therefore, it is hoped that the robust methods used in this study would contribute to MWI efforts in improving water management by applying remote sensing methods to uncover violations related to irrigational water use and to audit annual records of water abstraction. Adoption of these methods towards water accounting systems, however, would require capacity building in the use and utilization of remote sensing data and geospatial techniques.

5. CONCLUSIONS

Remote sensing data and geospatial techniques act as good sources and tools for providing data needed for managing the scarce water resources of drylands. The study showed that the improved techniques for identifying irrigated areas and crops using remote sensing would include the use of multi-temporal imagery and ancillary data. The contribution of geospatial techniques was mainly in water auditing as they provided crucial information on irrigated areas, cropping patterns and estimates of groundwater abstraction in relation to available water resources and irrigated crops. In terms of water use, the study showed intensive irrigation in the drylands of Jordan, particularly in Amman-Zarqa basin. At present, irrigation is consuming invaluable water resources and resulting in groundwater depletion. It can be concluded that inappropriate cropping patterns may threat scarce water resources of drylands. However, information provided through remote sensing and geospatial techniques may provide solutions that optimize cropping pattern and water use. This fact was recognized by decision makers in Jordan and amendments were made to the "Water Law" so that water accounting and auditing by means of these contemporary techniques were approved. The remaining challenge, however, will be the capacity building needed for adopting and implementing these techniques. Once this target is achieved, then it is hoped that remote sensing and geospatial techniques are taken one step further towards the development of water accounting systems for the different basins in Jordan.

ACKNOWLEDGMENTS

This publication and crop mapping work were supported by Ministry of Water and Irrigation (MWI), Jordan. The work was carried out through a project "Regional Coordination on Improved Water Resources Management and Capacity Building Project" funded by the Global Environment Facility (GEF), managed by the World Bank (Project ID P117170), and jointly implemented in Cooperation with NASA. The authors acknowledge the cooperation of MWI and its staff, particularly the Secretary General Eng. Basim Telfah and the Ministry Technical Assistant Eng. Ali Subh for their full support and cooperation. The authors appreciate the efforts and cooperation of Su'ad Nasser, Ali Hayjneh and Issa Al-Obra. The authors would like to thank partners from NASA, particularly Dr. Shahid Habib.

AUTHOR CONTRIBUTIONS

Jawad Al-Bakri worked on the methodology for crop mapping and NCWR comparisons. All authors were involved in calculations and comparisons made in the manuscript. Jawad Al-Bakri led the manuscript writing with contribution from other authors, who were also involved in field data collection and image processing stage.

CONFLICTS OF INTEREST

The authors declare no conflict of interest.

REFERENCES

1. Safriel, U.; Adeel, Z.; Niemeijer, D. Drylands Systems. In *Ecosystems and Human Wellbeing: Current State and Trends*; Hassan, R., Scholes, R., Ash, N., Eds.; Island Press: Washington, DC, USA, 2005; Volume 1, pp. 623–662.
2. IUCN Global Drylands Initiative. Available online: http://www.iucn.org/drylands (accessed on 25 August 2015).
3. Al-Bakri, J.T.; Salahat, M.; Suleiman, A.; Suifan, M.; Hamdan, M.R.; Khresat, S.; Kandakji, T. Impact of climate and land use changes on water and food security in Jordan: Implications for transcending "The Tragedy of the Commons". *Sustainability* **2013**, *5*, 724–748.
4. MWI (Ministry of Water and Irrigation, Jordan). *Water for Life: Jordan's Water Strategy, 2008–2022*; MWI: Amman, Jordan, 2009.
5. Al-Bakri, J.T.; Duqqah, M.; Brewer, T. Application of remote sensing and GIS for modeling and assessment of land use/cover change in Amman/Jordan. *J. Geogr. Inf. Syst.* **2013**, *5*, 509–519.
6. Department of Statistics, Jordan. Available online: http://www.dos.gov.jo (accessed on 20 September 2015).
7. UNHCR (The UN Refugee Agency). A Briefing Note on Syrian Refugees, 2013. Available online: http://www.unhcr.org/ (accessed on 20 September 2015).
8. DoS (Department of Statistics, Jordan). *Iraqis in Jordan: Their Number and Characteristics*; DoS, Fafo and UNFPA: Amman, Jordan, 2007.
9. Schacht, K.; Gönster, S.; Jüschke, E.; Chen, Y.; Tarchitzky, J.; Al-Bakri, J.; Al-Karablieh, E.; Marschner, B. Evaluation of soil sensitivity towards the irrigation with treated wastewater in the Jordan River region. *Water* **2011**, *3*, 1092–1111.

10. Ozdogan, M.; Yang, Y.; Allez, G.; Cervantes, C. Remote sensing of irrigated agriculture: Opportunities and challenges. *Remote Sens.* **2010**, *2*, 2274–2304.

11. Inglada, J.; Arias, M.; Tardy, B.; Hagolle, O.; Valero, S.; Morin, D.; Dedieu, G.; Sepulcre, G.; Bontemps, S.; Defourny, P.; Koetz, B. Assessment of an operational system for crop type map production using high temporal and spatial resolution satellite optical imagery. *Remote Sens.* **2015**,*7*, 12356–12379.

12. Bastiaanssen, W.G.; Karimi, P.; Rebelo, L.-M.; Duan, Z.; Senay, G.; Muthuwatte, L.; Smakhtin, V. Earth observation based assessment of the water production and water consumption of Nile Basin agro-ecosystems. *Remote Sens.* **2014**, *6*, 10306–10334.

13. Gowda, P.H.; Chávez, J.L.; Howell, T.A.; Marek, T.H.; New, L.L. Surface Energy balance based evapotranspiration mapping in the Texas High Plains. *Sensors* **2008**, *8*, 5186–5201.

14. Liou, Y.-A.; Kar, S.K. Evapotranspiration estimation with remote sensing and various surface energy balance algorithms—A Review. *Energies* **2014**, *7*, 2821–2849.

15. Godfrey, J.M.; Chalmers, K. *Water Accounting: International Approaches to Policy and Decision-Making*; Edward Elgar Publishing: Cheltenham, UK, 2012; p. 336.

16. Dost, R.; Bastidas; Obando, E.B.; Bastiaanssen, W. *Water Accounting through Remote Sensing*; A Background Report on Water Accounting Plus (WA+) in the Awash River Basin; Land and Water Division, Food and Agriculture Organization of the United Nations (FAO): Rome, Italy, 2013; p. 56.

17. Karimi, P.; Bastiaanssen, W.; Molden, D. Water Accounting Plus (WA+)—A water accounting procedure for complex river basins based on satellite measurements. *Hydrol. Earth Syst. Sci.* **2013**,*17*, 2459–2472.

18. Karimi, P.; Bastiaanssen, W.; Molden, D.; Cheema, M. Basin-wide water accounting using remote sensing data: The case of transboundary Indus Basin. *Hydrol. Earth Syst. Sci.* **2013**, *17*, 2473–2486.

19. Karimi, P.; Bastiaanssen, W.G.M.; Sood, A.; Hoogeveen, J.; Peiser, L.; Dost, R. Spatial evapotranspiration, rainfall and land use data in water accounting—Part 2: Reliability of water accounting results for policy decisions in the Awash basin. *Hydrol. Earth Syst. Sci. Discuss.* **2015**,*19*, 1125–1167.

20. Al-Bakri, J.T. *Crop Mapping and Validation of ALEXI-ET in Azraq and Mafraq Areas*; A Report for Regional Coordination on Improved Water Resources Management and Capacity Building; Ministry of Water and Irrigation: Amman, Jordan, 2015; pp. 19–31.

21. Al-Bakri, J.T. *Crop Mapping for Azraq and Ramtha Areas*; A Report for the Advanced on-job Training on the Use of Remote Sensing in Crop Mapping and Evapotranspiration; Ministry of Water and Irrigation: Amman, Jordan, 2014.

22. Vörösmarty, C.J. Global water assessment and potential contributions from Earth Systems Science.*Aqua. Sci.* **2002**, *64*, 328–351.

23. Bastiaanssen, W.G.M.; Molden, D.J.; Makin, I.W. Remote sensing for irrigated agriculture: Examples from research and possible applications. *Agric. Water Manag.* **2000**, *46*, 137–155.

24. GIZ (Gesellschaft für Internationale Zusammenarbeit). *The Azraq Dilemma: Past, Present and Future Groundwater Management*; German-Jordanian Programme "Management of Water Resources", MWI (Ministry of Water and Irrigation, Jordan): Amman, Jordan, 2012; pp. 6–12.

25. MWI (Ministry of Water and Irrigation, Jordan). *Water Budget for Year 2013*; Ministry of Water and Irrigation: Amman, Jordan, 2014; pp. 14–17.

26. Al-Wer, I. Zarqa River Rehabilitation and Sustainable Management. Master's Thesis, Department of Land and Water Resources Engineering, KTH Stockholm, Sweden, 2009; TRITA-LWR Degree Project. ISSN 1651-064X ISRN LWR-EX-09-32.

27. Abdulla, F.; Eshtawi, T.; Assaf, H. Assessment of the impact of potential climate Change on the water balance of semi-arid watershed. *Water Resour. Manag.* **2009**, *23*, 2051–2068.

28. Al-Bakri, J.T.; Suleiman, A.; Abdulla, F.; Ayad, J. Potential impacts of climate change on the rainfed agriculture of a semi-arid basin in Jordan. *Phys. Chem. Earth* **2011**, *35*, 125–134.

29. Al-Bakri, J.T.; Brown, L.; Dedalof, Z.; Berg, A.; Nickling, W.; Khresat, S.; Salahat, M.; Saoub, H. Modelling desertification risk in the north-west of Jordan using geospatial and remote sensing techniques. *Geomatics Nat. Hazards Risk* **2014**.

30. The Highland Water Forum: Dialogue for Sustainability. Available

online: https://highlandwaterforum.wordpress.com/ (accessed on 17 October 2015).

31. Jensen, J.R. *Introductory Digital Image Processing: A Remote Sensing Perspective*, 3rd ed.; Prentice Hall: Upper Saddle River, NJ, USA, 2006; pp. 337–406.

32. El-Magd, I.A.; Tanton, T.W. Improvements in land use mapping for irrigated agriculture from satellite sensor data using a multi-stage maximum likelihood classification. *Int. J. Remote Sens.* **2003**, *24*, 4197–4206.

33. Simonneaux, V.; Duchemin, B.; Helson, D.; Raki, E.; Olioso, A.; Chehbouni, A. The use of high-resolution image time series for crop classification and evapotranspiration estimate over an irrigated area in central Morocco. *Int. J. Remote Sens.* **2008**, *29*, 95–116.

34. Mclver, D.K.; Friedl, M.A. Estimating pixel-scale land cover classification confidence using non-parametric machine learning methods. *IEEE Trans. Geo-Sci. Remote Sens.* **2001**, *39*, 1959–1968.

35. Mclver, D.K.; Friedl, M.A. Using prior probabilities in decision-tree remotely sensed data. *Remote Sens. Environ.* **2002**, *81*, 253–261.

36. Gumma, M.K.; Thenkabail, P.S.; Nelson, A. Mapping irrigated areas using MODIS 250 meter time-series data: A study on Krishna River Basin (India). *Water* **2011**, *3*, 113–131.

37. Gumma, M.K.; Thenkabail, P.S.; Hideto, F.; Nelson, A.; Dheeravath, V.; Busia, D.; Rala, A. Mapping irrigated areas of Ghana using fusion of 30 m and 250 m resolution remote-sensing data. *Remote Sens.* **2011**, *3*, 816–835.

38. Zhu, Z.; Woodcock, C.E. Object-based cloud and cloud shadow detection in Landsat imagery.*Remote Sens. Environ.* **2012**, *118*, 83–94.

39. Smith, M.; Milton, E. The use of Empirical Line Method to Calibrate Remotely Sensed Data to Reflectance. *Int. J. Remote Sens.* **1999**, *20*, 2653–2662.

40. Du, Y.; Teillet, P.; Cihlar, J. Radiometric Normalization of Multitemporal High-resolution Satellite Images with Quality Control of Land Cover Change Detection. *Remote Sens. Environ.* **2002**, *82*, 123–134.

41. Jensen, J.R. *Introductory Digital Image Processing: A Remote Sensing Perspective*, 3rd ed.; Prentice Hall: Upper Saddle River, NJ, USA, 2005; pp. 337–406.

42. Pulighe, G.; Baiocchi, V.; Lupia, F. Horizontal accuracy assessment of very high resolution Google Earth images in the city of Rome, Italy. *Int. J. Digital Earth.* **2015**.

43. Eastman, J.R.; Filk, M. Long sequence time series evaluation using standardized principal components. *Photogramm. Eng. Remote Sens.* **1993**, *59*, 991–996.

44. Congalton, R.G. A review of assessing the accuracy of classifications of remotely sensed data.*Remote Sens. Environ.* **1991**, *37*, 35–46.

45. Bastiaanssen, W.; Menenti, M.; Feddes, R.; Holtslag, A. A remote sensing surface energy balance algorithm for land (SEBAL) 1. Formulation. *J. Hydrol.* **1998**, *212*, 198–212.

46. Bastiaanssen, W.; Pelgrum, H.; Wang, J.; Ma, Y.; Moreno, J.; Roerink, G.; van der Wal, T. A remote sensing surface energy balance algorithm for land (SEBAL) 2. Validation. *J. Hydrol.* **1998**, *212*, 213–229.

47. Allen, R.G.; Pereira, L.A.; Raes, D.; Smith, M. *Crop Evapotranspiration*; FAO Irrigation and Drainage Paper 56; Food and Agriculture Organization of the United Nations (FAO): Rome, Italy, 1998; p. 293.

48. Allen, R.G.; Pereira, L.A. Estimating crop coefficients from fraction of ground cover and height.*Irrig. Sci.* **2009**, *28*, 17–34.

49. FAO (Food and Agriculture Organization of UN), Land and Water Division: ETo Calculator. Available online: http://www.fao.org/nr/water/eto.html (accessed on 22 October 2015).

50. Abu-Awwad, A.; Blair, S. Economic Efficiency of Water Use by Irrigated Crops in Al'Azraq Area.*Jordan J. Agric. Sci.* **2013**, *9*, 525–543.

51. Mazahrih, N.T. Evapotranspiration Measurements and Modeling for Bermuda Grass, Alfalfa, Cucumber, and Tomato Grown under Protected Cultivation in the Central Jordan Valley. Ph.D. Thesis, The University of Jordan, Amman, Jordan, 2011.

52. Suwwan, M.; Judah, O. Influence of Plastic Mulching on Growth, yield and Soil Moisture Conservation in Plastic House Tomatoes. *Dirasat* **1985**, *12*, 21–32.

53. Al-Bakri, J.T.; Taylor, J.C.; Brewer, T.R. Monitoring land use change in the Badia transition zone in Jordan using aerial photography and satellite imagery. *Geogr. J.* **2001**, *167*, 248–262.

54. JTN (The Jordan Times). Satellite Remote Sensing Uncovers violations on Underground Water Resources. Available online: http://www.jordantimes.com/news/local/satellite-remote-sensing-uncovers-violations-underground-water-resources#sthash.f4mMEUaF.dpufhttp://www.jordantimes.com/news/local/satellite-remote-sensing-uncovers-violations-underground-water-resources(accessed on 25 October 2015).

55. Hamdan, I. *Hydrogeological and Hydrological Investigation for Azraq Basin—North East Jordan*; Deutsche Gesellschaft für Internationale Zusammenarbeit (GIZ): Amman, Jordan, 2010.

56. Campos, I.; Neale, C.M.U.; Calera, A.; Balbontín, C.; González-Piqueras, J. Assessing satellite-based basal crop coefficients for irrigated grapes (*Vitis vinifera L.*). *Agricul. Water Manag.* **2010**, *98*, 45–54.

57. Salman, A.Z.; Al-Karablieh, E. Measuring the willingness of farmers to pay for groundwater in the highland areas of Jordan. *Agricul. Water Manag.* **2004**, *68*, 61–76.

58. Margane, A.; Al-Qadi, M.; Al-Kurdi, O. *Updating the Groundwater Contour Map of the A7/B2 Aquifer in North Jordan*; Technical Cooperation Project 'Syrian Refugee Response', Technical Report No. 1; Bundesanstalt für Geowissenschaften und Rohstoffe (BGR) and Ministry of Water and Irrigation, Jordan (MWI): Amman, Jordan; p. 129.

59. El-Naqa, A.; Al-Momani, M.; Kilani, S.; Hammouri, N. Groundwater Deterioration of Shallow Groundwater Aquifers Due to Overexploitation in Northeast Jordan. *Clean* **2007**, *35*, 156–166.

60. Law Amending the Law of the Water Authority (In Arabic). Law number 22, 2014. Jordan Prime Ministry, Official Gazette: Amman, Jordan, 2014; pp. 3385–3390.

Index